CHINESE BUSINESS AND THE ASIAN CRISIS

Chinese business has been an emerging and distinctive current in global capitalism with a history and attributes different from those of the Japanese and the various kinds of Western business. This book provides an important demonstration that as well as losers in the Asian crisis, there have been survivors and new winners. A stimulating analysis of changing strategies and modifications to culture and structures, this book provides essential reading to business economists and especially to those researching the development of mainland China as a new global actor.

Chinese Business and the Asian Crisis

Edited by

DAVID IP
University of Queensland

CONSTANCE LEVER-TRACY
Flinders University of South Australia

NOEL TRACY
Flinders University of South Australia

Gower

© David Ip, Constance Lever-Tracy and Noel Tracy 2000

Published by
Gower Publishing Limited
Gower House
Croft Road
Aldershot
Hampshire GU11 3HR
England

Gower
131 Main Street
Burlington
Vermont 05401
USA

British Library Cataloguing in Publication Data
Chinese business and the Asian crisis. - (Explorations in
    Asia Pacific business economics)
    1.Business enterprises - China - Finance 2.Financial crises
    - Asia 3.China - Economic conditions - 1945- 4.Asia -
    Economic conditions - 1976-
    I.Ip, David II.Lever-Tracy, Constance III.Tracy, Noel
    338.5'42'095

Library of Congress Control Number: 00-132845

ISBN 0 7546 1342 9

Printed and bound by Athenaeum Press, Ltd.,
Gateshead, Tyne & Wear.

Contents

List of Tables and Figures

viii

List of Contributors

The Editors

David Ip (PhD) is currently Associate Professor in the Department of Anthropology, Sociology and Archaeology at the University of Queensland and the Director of the Master's program in Social Planning and Development. He was born in Hong Kong and educated there as well as in the United States and Canada. His main research interests are in Chinese immigrant entrepreneurs in Australia on which he has published widely. He is co-author of *Impressions of Multicultural Australia*. He is also actively involved in consultancy work for the Australian international development assistance bureau, AusAID and other international development organisations.

Constance Lever-Tracy (PhD) is a senior lecturer in the Department of Sociology and in the School of Political and International Studies, Flinders University of South Australia, Adelaide. She teaches and supervises student dissertations in the University's Masters programme in International Relations, in Hong Kong and China. She was educated at the London School of Economics, University College London and Flinders University. She has a long standing interest in international migration and in globalisation on which she has published widely and is co-author of *A Divided Working Class? Ethnic Segmentation and Industrial Conflict in Australia*.

Noel Tracy is Associate Professor in the School of Political and International Studies, Flinders University of South Australia, Adelaide and coordinator of the Masters in International Relations overseas programmes. He was educated at La Trobe University, Melbourne and at Flinders University. His main research interests are in the political economy of industrialisation and economic transformation of China on which he has published widely. He is author of *China's Export Miracle* (1999).

The three editors have researched together for over a decade. Their joint publications include *Asian Entrepreneurs in Australia* (1991) and *The Chinese Diaspora and Mainland China: An Emerging Economic Synergy* (1996).

The Contributors

Cen Huang (PhD) is the Director of International Programs and Partnerships in the University of Calgary. She was a research fellow at the International Institute for Asian Studies in Leiden from 1996-1999. Her research work is concerned with the structure and social organization of overseas Chinese invested enterprises in South China. She has done extensive fieldwork in South China and Southeast Asia in the past three years and has published widely on the research topic. Dr. Huang has also been involved in cross-cultural management of joint foreign ventures and educational reforms in China.

Henry Wai-chung Yeung (PhD) is Assistant Professor at the Department of Geography, National University of Singapore. He is a recipient of NUS Outstanding University Researcher Award 1998 and Institute of British Geographers Economic Geography Research Group Best Published Paper Award 1998. His research interests cover broadly theories and the geography of transnational corporations, Asian firms and their overseas operations and Chinese business networks in the Asia-Pacific region. Dr. Yeung has published widely on transnational corporations from developing countries, in particular Hong Kong, Singapore and other Asian Newly Industrialised Economies. He is the author of *Transnational Corporations and Business Networks: Hong Kong Firms in the ASEAN Region* (1998), editor of *The Globalization of Business Firms from Emerging Economies* (1999) and co-editor of *Globalisation and the Asia Pacific: Contested Territories* (1999) and *Globalization of Chinese Business Firms* (2000).

Acknowledgments

The editors wish to thank the Australia Research Council for a grant to themselves and to associate researcher Dr. Zhu Wenhui, to study the Chinese diaspora and the Asian crisis. This project is still in progress, but its preliminary findings have contributed significantly to this book, in particular to Chapters 6, 8 and 9.

Chapter 4 is a revised version of a paper that appeared in September 1999 in the *Bulletin of Concerned Asian Scholars*. The author is grateful for permission of the journal editors to make use of this material.

Chapter 5 is a revised version of an earlier paper presented at a workshop on *Challenges to the Chinese Overseas in an Era of Financial Vulnerability in the Asia-Pacific Region*, University of New South Wales, Sydney, Australia, 14-15 May 1999 and at seminars at the University of Sydney and the University of Bristol. Comments from participants at various occasions, in particular Connie Cheng, David Demeritt, Constance Lever-Tracy, Andy Leyshon, Liu Hong, Terry McGee, Nigel Thrift and Noel Tracy, are gratefully acknowledged. The collection of materials presented in this paper is funded by the National University of Singapore (RP960045 and RP970013). The author would like to thank Wan Meng Hao and Elen Sia for their excellent research assistance. The paper has previously appeared in the Centre for Advanced Studies Research Papers Series, No.14, National University of Singapore, 1999.

The author of Chapter 6 and Chapter 8 wishes to acknowledge Richard Lam and Richard Hsu for their inputs to the interviews and IIAS (International Institute for Asian Studies) in Amsterdam for providing institutional support during the writing of this paper.

1 Introduction

DAVID IP, CONSTANCE LEVER-TRACY AND NOEL TRACY

From Crisis to Recovery

In January 1997 Hanbo Steel, a large Korean chaebol, collapsed under $6 billion in debts. In mid May, Thailand's baht currency was hit by a massive attack by speculators. Thereafter the crisis deepened and spread rapidly throughout the region. A panic flight of short-term capital revealed and also induced a mountain of bad debts, bringing currencies and economic activity crashing down. By the end of 1998, there was still no evidence of sustained recovery and it seemed clear that IMF rescue regimes had failed or made the situation worse. What had started as a financial and currency crisis had devastated the real economy of the region, as the costs of imported food and necessary inputs for production and of debt servicing and repayment soared and as new credit for productive activities dried up or became phenomenally expensive (Krugman, 1998; Sachs, 1997). Poverty and unemployment increased rapidly, and in that year throughout East and Southeast Asia, only China and Taiwan and (barely) Singapore avoided negative growth (*FEER*, 27 January, 2000, pp. 80, 81).

A year later, as the new century dawns, the memory of the Asian Financial Crisis is fading around the region, like a bad dream. Economic growth, throughout much of non-Japan Asia, has been recovering fast, boosted in part by rising exports to Europe and America and in part by reviving domestic demand and intra-regional trade (*FEER*, 1 July, 1999, p. 51; 5 August, 1999 p. 41). Stock markets and property values have been soaring again; technology companies have been listing on the Nasdaq; e-commerce and hi-tech start ups are proliferating, and unemployment is in decline (*FEER*, December 1999, January 2000, passim). The banking systems are indeed still trying to cope with the accumulated mountain of bad debts, but although banks remained reluctant to lend, firms were learning how to live without them and to find other, perhaps less short term And volatile sources of finance (*FEER*, 3 June, 1999, p. 53; 29 July, 1999 pp. 30-35).

The speed of recovery, once it got going, had been unexpected even by the optimists, and estimates and projections of growth had to be revised upwards substantially several times (see Table 1.1). Yet again events in Asia had taken experts and theorists of diverse standpoints by surprise, as indeed had the previous crisis and before that the 'economic miracle' of rapid development as well.

During the previous year, confident projections of doom had become common. In August 1998, for example, the Asia Research Centre at Murdoch University (Perth, Western Australia) organised a conference with the title *From Miracle to Meltdown: The End of Asian Capitalism,* without even the precaution of a question mark. Chris Patton's BBC documentary *East Meets West* saw the prior successes as a bubble now pricked. There was a new orthodoxy among business commentators. Recovery (if possible at all) would require a 'new economic order', total restructuring and an end to 'crony capitalism', to be achieved by the intervention of bodies such as the IMF and by Western buyouts, leading to a complete overhaul of Asian management. In October the *Far Eastern Economic Review* (29 October, 1998) was advising investors to put their money 'under the bed' (or into US government bonds). Some academics such as Bello and Rosenfeld (1990) had long argued that the 'Asian miracle' was but a mirage based on an 'ersatz capitalism'. Although the pre-crisis long term development in the region had marginalised this view, it now, for a time, came into its own again.

As the renewed growth took hold, some continued in denial and others expressed pique, complaining that it had not waited for the transformations they had decreed to be essential. (*FEER,* 13 May, 1999, p. 47; 1 July, 1999 p. 42; 19 August, 1999 p.30; 7 October, 1999, p. 87). A 'premature' recovery would forestall 'the will to modernise' (*FEER* 17 June, 1999, pp. 38-40). Returning investors were abused as 'lemmings with the attention span of a mosquito' (*FEER,* 13 May, 1999, p. 47) for ignoring that new bankruptcy laws had scarcely been used, companies were dragging their feet on debt settlement, Western companies had been blocked from acquiring control of important companies, 'openness' and 'transparency' were little advanced and family management and personal networks were still prevalent. Alas, they could not stop the rising tide, it seemed the markets did not care about the experts' reforms (*FEER,* 17 June,1999, p. 53).

Reality checks had not been kind to the optimists either. As the unexpected crisis struck, many who had celebrated the earlier economic successes had stressed again the soundness of 'economic fundamentals' and of the 'real economies' of the region and suggested the flight of hot money and the collapse of currencies would be temporary and superficial problems (eg Hamilton, 1998). Sadly, a slight move upwards in early 1998 had then proved to be illusory (*FEER,* 23 April, 1998, p. 34) and the impact of the crisis on the productive capacity and the lives of people in the region was devastating for nearly two years. When the storm had passed, it was not just a case of continuing as before. The landscape had changed and there were new winners as well as long-term losers.

There have been contradictory explanations for the crisis, and from these have flowed opposing prescriptions. One side has stressed internal problems of corruption and 'moral hazard', induced by government favours and a system thriving on opacity. Remedies advocated focussed on reimposed market discipline, through withdrawal of government protection and subsidies, enforced transparency and the encouragement of bankruptcies and foreign buyouts. The other side has pointed to markets out of control, in particular the weak regulation of rising tides of international hot money flows, seeking short term profit and liable to vicious circles of ignorant panic. They have advocated capital controls and positive government. (See Chapter 4).

Yet hindsight has not brought resolution and we can find no clear vindication of any of the remedies. Malaysia and Thailand have both recovered at roughly comparable rates, although the first introduced capital controls and the second tried to follow IMF prescriptions (*FEER,* 8 July, 1988, p. 82). Hong Kong and Singapore have both been praised as havens of sound banking and incorruptibility. The former has been one of the slowest to recover, the latter one of the fastest. Bodies such as the IMF and the World Bank have themselves expressed doubts, in retrospect, about their own previous prescriptions (*FEER,* 7 October, 1999, p. 86; 28 October, 1999, p. 77).

If the recovery has thrown the explanations of the crisis up in the air, the crisis in turn had thrown into question the diverse explanations for the preceding economic success of the region. Where some had claimed the credit for strong developmental state policies others had asserted the primacy of free global markets. Although each side, as we have seen, tried to blame the other for the turnaround, it was not clear what had so rapidly upended the putative balance of forces or led to the apparent degeneration of the claimed positive drive.

Each of the three phases, growth, crisis and recovery, casts doubt on the adequacy of explanations credibly propounded for the prior phase. A closer look reveals also a synchronous diversity. No overall theory is satisfactory for all phases or all places. Such theories have indeed often acted as blinkers, encouraging too narrow a focus on the factors central to theory - the policies of national governments and international agencies, the actions of Western or Japanese corporations, the ebb and flow of global markets - to the exclusion of others, particularly of the rising capitalist class within the region.

Table 1.1 Revisions to Economic Growth Forecasts[1]

	For 1999		For 2000	
Date of forecast	Mid 1999	End 1999	Mid 1999	End 1999
Singapore	2.5	5.0	4.0	5.6
Malaysia	2.0	5.2	3.9	5.0
Indonesia	-1.8	-0.4	2.4	2.4
Thailand	-0.5	3.7	2.2	4.0
Philippines	2.4	3.0	4.0	4.0
Hong Kong	-1.0	2.0	1.5	4.2
Taiwan	4.5	5.3	4.8	5.5
China	8.2	7.3	8.4	7.6
Japan	-0.7	0.8	0.9	0.0

Sources: *FEER*, 1 June, 1999, pp. 54,55; *FEER*, 27 January, 2000, pp. 80,81 (based on Goldman Sachs, Asia).

This book aims to bring this rising class into central focus, not only the more visible tycoon conglomerates but also the vast mass of small and medium family firms. These capitalists emerged as a new and distinctive current within global capitalism, capable of an exceptional dynamism, during the decades of the 'Asian economic miracle' and the opening of Mainland China. They were able to establish a variety of regional and global linkages, but the crisis also revealed mismatches at the interface between them and other capitalist currents and between their different ways of operating. The recovery, however, is also demonstrating their capacity to survive and adapt - Asian capitalism is still alive and kicking.

Ethnic Chinese businesses have been and are still the most important, dynamic and externally oriented part of this local private capitalism within both East and Southeast Asia, and they are responsible for the lion's share of foreign investment in Mainland China, serving as its bridge to the global economy. While there are other important indigenous and migratory business groups within the region, there can be no doubt about the primary role in local capitalist enterprise of the over 50 million diaspora Chinese, who now live in Hong Kong, Taiwan and Southeast Asia, with linkages to smaller communities in Australia, the Americas and most other countries around the world. It is on these, the impact on them of the Asian Financial Crisis of 1997 and their responses to it, that this book will concentrate.

Chinese Diaspora Capitalism

We take diaspora to be a collective noun, referring to people who have (themselves or their ancestors) been scattered from a common place of origin, and to the elements of shared culture and inter-relationships that continue to link them. Insofar as a diaspora is constituted by shared memories and common attributes, these are likely, with the passage of time, to fade and assimilate as it adapts to its diverse environments. Its survival and rejuvenation depend on continuing transnational interconnections and new movements as well as the making for its members of a distinctive position in its various places of settlement. Thus, historically, the most tenacious diasporas have often been those linked both to their local societies and to other parts of the diaspora by long distance trade, who can combine 'ethnic identity and cosmopolitan adaptability' (Kotkin, 1993). Chinese have been important diaspora traders around the South China Sea since well before the 12th century and they were the predominant middlemen in precolonial and colonial times in Southeast Asia. While Western capitalism developed within the framework, and with the support of rising nation states, the Chinese diaspora established a 'peripheral capitalism', on the fringes or outside the reach of the Imperial and Communist states. Most came from the outer parts of the empire, from its southern coastal provinces, where state control was less tight, or via the treaty ports or the ceded territories of Hong Kong or Macao. Their communities around the region and the world have largely established themselves, and their relations with each other, as self regulating entities, without sponsorship or protection by the Chinese state and often by evading its law, in sharp contrast with most colonists, merchants and multinationals from Europe or America.

They created their own distinctive institutions by reshaping the traditions they had brought with them and creating new ones, and they adopted local practices and beliefs and married local women. Yet the links with each other and with mainland China were never fully broken for long, for such links constituted their special advantage, and the diaspora was periodically replenished by cross migration and new waves.

China has for over two thousand years been a great power in the east yet it has nearly always seen itself as a land based empire, indifferent or actively hostile to the traders and emigrants who left its shores. Chinese merchants led the way in East and Southeast Asia by the 12th century and dominated regional trade by the 14th century. Between 1400 and 1430 there was a wave of state sponsored maritime explorations, but these were soon sharply cut off and, until the second half of the 16th century, international trade itself became illegal and thereafter tightly restricted. In the mid-17th century the emperor forbade emigration and in the following century those resident overseas were ordered home, although many disobeyed. The new waves of departures from southern China in the 19th and early 20th century were at first overwhelmingly male, as women were for long prevented from leaving. After 1949 the Communists again sought to close the borders, generally more successfully than their predecessors, although there was a continuing influx of illegal emigrants to Hong Kong (Wang, 1991).

Another concentration of sometimes unlawful emigrants from the mainland was based in Taiwan, which was only attached to China late and insecurely. Settled from China from the 16th century, often by merchants and pirates from Fujian, who were seeking a base for maritime activities, and by rebels against the Manchu dynasty, it only came under central control after 1684, but was always lightly administered and notoriously lawless. From 1895 until 1945 it was a colony of Japan and since 1949 has been ruled by KMT exiles from mainland Communism (Wang, 1991).

A paradox, of continuity and discontinuity, characterises these communities. The existence of a Chinese diaspora is, in many places, older than both the postcolonial nations and the colonial states that preceded them. Yet its notables today are largely new men and women, immigrants who have worked their way up from coolies and plantation workers or from shopkeepers and small commodity traders to riches, or they are the children of such immigrants. On the one hand, one can find direct continuities, in diaspora business culture today, with features of the merchant culture and institutions in the China of 500 years ago. On the other hand, most of the members of these communities can trace their own family's origins to villages in China only a couple of generations ago. For most in Southeast Asia the migration occurred in the period preceding the last war. For most in Hong Kong it is even more recent - until 1981 those born in China were a majority. The explanation lies in the recurring new migratory waves, whose thrusting and ambitious members have first worked for or coexisted with the established and learnt from them, and then often displaced them, as the *Totok* displaced the *Peranakan* in Indonesia and the *Babas* in Malaysia. Many of the longer established have, over the generations, assimilated into the local society, and some have ossified into economically declining, tradition-bound sub communities, while newcomers have taken their place. A persisting feature of these ancient communities then is their intense social mobility and constant self rejuvenation[2].

One long standing culturalist explanation, for the survival and success of Chinese business around the world, located the prime cause in a continuing tradition of Confucian ethics which was said to promote hard work, discipline and entrepreneurship (e.g. Hofheinz and Calder, 1982). There are many problems with this approach, which does not explain the absence of capitalist transitions within the mainland over the centuries. Confucian ideas are indeed non fatalist and this worldly, valuing education and self improvement, and stressing personal morality. They also, however, include strong inhibitors of capitalist development, especially in their contempt for trade, their disapproval of risk and profit and their stress on ritual and hierarchy. Wang shows how these ideas have had to be selectively reshaped before they could be of use to traders or entrepreneurs, rather than an obstacle in their path. (Wang, 1991, Ch. 14).

Lam et al. call many Confucian ideas 'stultifying' and suggest that in Taiwan they serve as the ideology of Taiwan's large bureaucratic organisations, not of its myriad, small dynamic entrepreneurs. These latter have been more inspired, they claim, by the underground anti authoritarian counter cultures of Taoist and Zen Buddhist heterodoxies, that for long have flourished on the Chinese periphery. These ideas encourage them to behave as 'rebels and bandits', and to 'challenge segments of industry dominated by... large scale enterprises, question the established order of brand name products and pirate know how and technology' (1994, p. 211).

Despite much that had to be jettisoned, migrating merchants and peasants both did bring valuable traditional resources with them, which they were able to develop into their own distinctive and autonomous institutions and modes of operating. As well as heterodox beliefs these included a cultural and ritual underpinning to monetary transactions, a multiplicity of associations, reliable networks and strong but flexible family structures and values which were the basis of family firms (Wang, 1991, Ch. 10).

Freedman, in a famous essay (1979), wrote of the sophistication of Chinese peasants in dealing with money. For small loans peasants turned to each other, to kinsmen or neighbours. Unlike the European Middle Ages, moneylenders were not a distinct class but everyone who had a surplus, and he cites an observer as saying that 'The whole Chinese empire may be said to be in a perpetual state of borrowing and lending' (p23). Mutual credit organisations were common, supported by ritual and based on trust (p25).

Sangren (1984), in a paper titled 'Traditional Chinese Corporations: Beyond Kinship', writes that 'The construction of formal groups on the basis of seemingly interchangeable criteria of membership is a facet of Chinese society noted most frequently by observers of overseas Chinese society'. He demonstrates the rural precedents within China for this Chinese 'virtuosity in forming co-operative organisations' (p. 410) on the basis of opportunistically shifting recruitment criteria that combine ascriptive and chosen ties (kinship, lineage, common surname, shared place of origin or dialect, religion etc). He concludes that kinship was but one possible basis for 'a cultural system clearly capable of generating a much greater range of creative organisational responses to changing historical and environmental circumstances' (p. 411).

Hamilton (1991) contrasts Chinese and European guild traditions in ways that emphasise the greater communitarianism of the former. It was merchant associations in Imperial China, not the state, that guaranteed creditable market behaviour by setting standards for weights and measures, enforcing contracts and supporting credit institutions. Since particular urban trades were in the hands of migrants from particular rural localities, these associations were also based on shared place of origin. 'Collegial relations formed the foundation for merchant and artisan associations; kinship relations formed the foundations for family firms' (p. 60).

The diaspora minorities around the world, and the settlers in Taiwan, responding to circumstances and on the basis of these foundations, constructed strong and durable commercial institutions and cultures which enabled them to raise credit, gather information and enter securely into agreements, without depending on states or legal systems. They formed a multiplicity of associations and they carried out their business through long term particularistic networks on the basis of personal trust, guaranteed by the indispensability of reputation within their communities and ostracism of those who forfeited it[3].

Diasporas in the Eras of Nationalism and of Globalisation

Much has been written about the importance of imperialism for dominant capitalisms and of strong states for late developers such as Germany, Japan or Korea (Gerschenkron 1962). It would seem that Imperial China forfeited its chance for global power when it turned its back on the world trading system in the 15th century. Conversely, its scattered trading communities were also severely handicapped in the world capitalist system, in the absence of such state support, during the colonial and nationalist eras.

Where Chinese have formed minority communities (in Southeast Asia, or in the Americas, Europe, Australia and elsewhere) they have faced states that were at best supportive in limited, discontinuous and unreliable ways and at worst actively hostile (Light, 1972; Cushman and Wang, 1988; Jesudasen, 1989; Chirot and Reid, 1997). Colonial states may have sponsored them as middlemen to the indigenous population, under the hegemony of the colonisers' capital, but at the same time subjected them to numerous restrictions.

In postcolonial times, individual Chinese have often found ways to form alliances with powerful persons in the local political elites or the military, as 'cronies', *cukongs* or proteges of 'guardian angels', but these have been as unstable as the position of the powerful officeholder. While the arrangement has provided favours for particular Chinese families it has not provided general state support for their business community as a whole. On the contrary, there have been numerous policies in the new Southeast Asian states aiming at their replacement as a business community by indigenous enterprise (Suryadinata, 1988; Kano, 1989; Mackie, 1988, 1989; Suehiro, 1989). Under both colonial and post-colonial regimes they have experienced periodic persecution and expulsion[4].

A similar distance or hostility of the state applied, to a degree, even in the majority Chinese societies of Hong Kong, Taiwan and Singapore. Hong Kong was a British colony with a traditional gulf between government and indigenous business, where until the 1970s the civil service was largely run by expatriate Britons and the heights of the economy were in the hands of British companies. While the state did provide public order and the rule of law as well as substantial infrastructure, public housing and education, it adopted a hands-off *laissez faire* approach towards the economy, long before this was popular elsewhere, and explicitly disavowed any role of supporting local business or furthering its interests (Chiu, 1988).

After Japanese colonial rule, between 1895 and 1945, Taiwan was taken over by mainlanders, KMT refugees from Communism, with little input by the local population, who were indeed significantly alienated from the state by the massacre of 1947. It has been described as a state with exceptional relative autonomy from both local landowners and capitalists, whose government (persuaded by modernisation theories) held attitudes unfriendly to small business, and where neither state banks nor the law were seen as serving them well. Since the 1970s, in particular, this state has had little standing in the international community (Wang, 1990, p. 265; Greenhalgh,1988; Evans and Pang,1989, p. 5; Kao 1991; Orru, Biggart and Hamilton, 1997). Independent Singapore, despite its majority Chinese population, is of the three the one where non-Chinese foreign capital has continued to have by far the most weight and influence with a government that for long saw them as the agents of modernisation. The Singapore state is, in any case, also a weak force in the international arena.

The 1994 edition of *The Dictionary of Sociology* still had no entry for 'diaspora'. For long diasporas and trading minorities were considered of little relevance in the modern world of nation state societies, at best serving a transitional role, soon doomed to disappear. Indeed, until recent decades, their place in the economy of the region and the world had seemed marginal and declining. Asia had been divided by colonialism and then by cold war boundaries and by autarkic national development policies. There were relatively few cross-flows of trade and investment between the countries of the region or between East and Southeast Asia or into China, and many potential social and economic networks lay dormant. Chinese communities increasingly diverged from each other.

In the last quarter of the 20th Century, however, both the reality and the concept of globalisation have come to the fore. In recent decades, trans-border flows of trade, investment, people and information have grown massively both globally and within the region. This new world situation has provided novel opportunities for a diaspora capitalism that had an established trading and business competence and transnational linkages that could be activated. The new flows often followed and reopened the old paths of now reviving Chinese networks. Today's technology and globalising economy could develop and enhance these resources and thus empower such a diaspora to aspire to be a challenger on the global stage.

The law of combined and uneven development suggested that any contender for capitalist power status would have to begin its industrialisation using the same levels of technology as the existing powers. We can apply the same reasoning to its transnational capacity. Capitalism is increasingly globalised and its main players are multinational corporations. In this situation any would-be competitor must start at the same global level and be able to start operating transnationally very quickly. Nor are all the cards in the hands of the established, bureaucratic, multinational corporations. The processes of globalisation, which multiply competition and unpredictability, have not always been comfortable for them. Their historic failure to expand capitalism into the old autarkic Communist states of the ex Soviet Union and their very minor role as investors in China have been indicators of timidity and limited competence in the new world situation.

Diaspora capitalists started with a transnational network and therefore had overcome obstacles to transnational operations before they started, and could not be frozen out of global marketing channels by existing powerful corporations or their state backers. They were accustomed to insecurity and to operating in unfamiliar environments, relying on family management of transnationally networked small and medium-sized firms, on personal trust and reputation and on strategies of flexible diversification, rather than on bureaucratic structures, law and state support or on the advantages of scale and mass production.

Many of the disorienting experiences of globalisation had long been familiar to the cultures of diasporas. In its modern form, however, globalisation has provided them with new resources for overcoming them. Through migration and remigration and long separations from family and friends they have always had to learn to foster relations with absent others, to live in locales thoroughly penetrated by distant influences, to accept what Giddens calls 'disembedding', the 'lifting out of social relations from local contexts of interaction and their restructuring across indefinite spans of time-space' (Giddens, 1990, pp. 17, 21). Now, however, through the telephone, the fax machine and the internet and taking advantage of cheapening air fares, they have been able to re-embed them and reconstitute a real and effective face to face community out of their scattered fragments around the globe, and one that can be called upon instantly for information and resources. As a minority in other people's territories, living in the cosmopolises of every age, they have always had to defend their identity against the danger Hannah Arendt saw in 'a global present without a common past [which may] threaten to render all traditions and all particular past histories irrelevant' (Hannah Arendt, cited in Robertson, 1992, p. 49). But now they could recover their particular histories, make pilgrimage to their place of origin, and to ethnic and religious shrines, organise international reunions for extended families or for old classmates and receive books and journals and videos from world centres of their languages and cultures.

The most sweeping claims for the central importance and aptitude of diasporas for the circumstances of the last years of the 20th century was made by Joel Kotkin. He called them 'global tribes', and defined them by their combination of ethnic identity, geographical dispersion (with a global network) and open mindedness (with a belief in scientific progress). He predicted that in the new era with the end of the Cold War and the reduced power of nation states, such cosmopolitan groups would flourish (Kotkin, 1993).

The Chinese diaspora did indeed develop into a significant new and distinctive current in global capitalism during the last quarter of the 20th century. It has been mainly in this period that their businesses have diversified from a concentration in trade to an important role in finance and manufacturing and from a status as middlemen, between local producers and imperialist economic powers, to that of an autonomous local capitalist class and increasingly a global actor. They acquired the most modern technical and business expertise but also found new strengths and relevance in their long established transnational connections and practices and the elaboration of these, in the new era of globalisation. Their centres in Hong Kong and Singapore won the status of regional and increasingly of world cities.

A Distinctive Capitalist Current

The Chinese diaspora has a multiplicity of national, political and class identities and loyalties and a diversity of cultural and historical roots. Yet the studies of Chinese economic activities around the world indicate, over and above the differences, a shared tendency to concentrate in (most often small) business activities with certain common features. This has made it possible to speak of a distinct type of capitalism, with particular strengths and weaknesses.

Independent business activities have provided the predominant role model, the community leadership and often the most common activity for mature adults in diaspora communities. In traditional China successful traders had sought to elevate the family's status by buying an education for their sons that would take them out of the merchant class and into the mandarin state bureaucracy. In the diaspora the outsider status of the Chinese tended to exclude this possibility (along with other positions in agriculture or the military for example), and discrimination often also closed off to them the managerial levels of Western companies. Within their own business community, managerial and bureaucratic positions were rare and, where they existed, tended to be subject to owners' mistrust, while access to top positions was reserved for family members. Both opportunities and blockages pushed people with any ambition into self employment. Becoming your own boss was a widespread aspiration (frequently achieved). Even education has been more likely to lead back (and feed back) into business, or at least into an independent profession, than into high status employment. The 'fetters on gain' that Weber claimed had applied within Imperial China, had been rejected.

The features that have distinguished diaspora Chinese capitalism from the mainstream of Western or Japanese forms of capitalism have, arguably, been more of degree than of kind, more a case of the greater frequency and legitimacy of certain attributes than a contrast of presence and absence. Nonetheless, in combination they add up to a characteristic business culture and mode of operation, with its own needs, strengths and weaknesses[5].

The central attributes include a persistence of family control over ownership and entrepreneurial decision making, even in the largest companies, where professional management and public flotation are well established. The tendency, with a few noteworthy exceptions, has been towards a multiplication of relatively small units in a conglomerate structure under the family's control, rather than the expansion of size and market share of large bureaucratically organised firms. This too has reduced the visibility of the concentrations of capital involved.

In its external relations (with lenders, borrowers, suppliers, customers, contractors and subcontractors etc), Chinese capitalism is distinguished by a preference for long-term but extensible, personalised networks, based on trust and upheld by the indispensability of reputation within such a system. Because of the nature and history of a diaspora, such networks are frequently transnational in scope, providing even small operators with reliable information and partners in other countries, around the region and indeed the globe. Where such connections are absent but desired, they can often be deliberately cultivated after personal introductions have been made.

A third feature of Chinese capitalism, at least in recent times, has been a preference for a strategy of diversification, in the interests of maximising flexibility, avoiding 'putting all your eggs in one basket' and taking advantage of novel and unpredictable opportunities that networks may throw up. This is also encouraged by the desire to allow individual family members to carve out a territory of their own, and is facilitated by the freedom of owner managers to make rapid decisions.

Western and Japanese systems of capitalism have tended to present a duality of large corporations and small and medium firms, with major differences between them and limited opportunities to move back and forth. In Chinese capitalism these features are common to both large and small operators, leading to greater similarities and continuities and opportunities for mobility up and down. Small firms, with large entrepreneurial ambitions and transnational networks, and with a leading role for professionally educated family members, can grow fast by multiplication. Tycoons retain the personalistic style of small operators and may be weakened or have their wealth split up if key managers leave to set up independently or if inheritance is divided.

What probably thus most distinguishes Chinese business from the mainstream forms that have been most powerful in North America, Western Europe and Japan is the predominance of entrepreneurship over organisation. The owner/manager, dealing through a series of well-established personal networks, is able to make deals rapidly and at a distance. Default is discouraged by fear of the loss of reputation for trustworthiness. We should note, however, that the Siamese twin of entrepreneurial risk taking is a taste for gambling and with it for speculation (Oxfeld, 1993; Ho, 1997).

Networks expand and contract, with some parts becoming dormant, according to the opportunities available. Chinese business needs to be seen as a series of networks, some of which interconnect at various points, rather than a single network. These interconnections allow members to be introduced into other networks, and thus expand and diversify their activities. Networking represents an alternative method of operating in a society in which the institutional arrangements do not necessarily support your activities or interests, and is a particularly apposite way for minorities to operate. Chinese business is, therefore, distinct in its operational and organisational form from the business structures of other major capitalist groups. Yet in this current period, its features can be seen as having analogies with the novel, rising forms of Italy's industrial districts or the high tech start ups of Silicon Valley, which in the 21st century are beginning to challenge the old order.

The Future

Chinese diaspora capitalism has a history older than that of Western capitalism. Decried by writers such as Max Weber as a 'traditional capitalism', lacking the potential of modern 'rational' forms, it was indeed largely marginal and restricted until the current wave of globalisation. Now it came into its own and showed itself well capable of competing and expanding. Yet one might ask whether their relish for the unpredictable tides of the global economy has not taken them sometimes out of their depth. Their reliance on personal trust has not always been adequate in the face of the bubbles and panics of hot money flows and their own culture of opacity, essential for family control and network confidentiality, has been a source of mistrust for Western fund managers and banks used to anonymous but transparent transactions.

Now the crisis has ebbed, the important questions are about its longer term impact. Will diaspora business draw in its horns and retreat to more modest ambitions? Will it simply adapt and converge to Western norms? Will it find ways of developing its own resources and competencies to forestall or prepare for recurrence? Will the outcome of the crisis be a vindication, a modification or an abandonment of the distinctive features of diaspora Chinese capitalism, in particular the autonomy of family firms, the use of networks and the strategies of diversification and transnationalisation of activities? What changes are to be found in the status and ranking of different parts of the diaspora and of different firms and sectors? Will Chinese diaspora capitalism overall be weakened vis a vis other global capitalist currents or might it emerge wiser and perhaps stronger?

The Book

The Asian Financial Crisis and its aftermath provide a crucible in which Chinese diaspora capitalism has been tested, and a new prism through which its strengths and weaknesses may be seen in a different light. The papers collected here are in many ways still tentative. Some represent work in progress reports on as yet uncompleted research. In many cases outcomes explored are still unclear or have not even yet fully unfolded. The aim is to focus on the consequences for diaspora Chinese capitalists and to start trying to identify losers and winners in the new landscape. We hope to start a re-evaluation of their business culture, strategies and modes of operation and of their likely future direction and potential.

Chapter 2, 'Chinese Diaspora Capitalism on the Eve of the Crisis', by Constance Lever-Tracy, sets the scene. This chapter discusses the real and substantial economic achievments of the 'Asian miracle' and the role of Chinese business in increasingly knitting together a regional economy and creating a bridge between China and the world market. It notes how academic understanding was hindered by the repetitious debate between 'strong states' and 'free markets', with a country by country frame of reference and an indifference to the actual class actors involved.

Commentators tended to be mesmerised by the presumed power of large, bureaucratic corporations, and Chinese business was dismissed as small and weak or as backward and traditional, even when contrary evidence accumulated. The chapter provides an overview of economic growth, regional interlinkages through trade and investment and the increasing autonomy and rising skills and technology of local actors. It concludes that academic theories tended to underestimate the potential of Chinese business attributes and the way they had been made increasingly relevant by the changing nature of a globalising world economy and its new technology.

In Chapter 3, 'The East Asian Financial Crash: Causes and Consequences', Noel Tracy places the Asian experience in the context of the history of speculative bubbles and busts, particularly that of 1929. He points to problems and weaknesses in international financial markets that led to the debacle. An excess of inadequately controlled global liquidity and a shortage of profitable outlets fueled a flood of 'hot money' pursuing short term profits, leading to massive overlending and overborrowing. When panic set in, the flight of capital soon became a flood and liquidity disappeared. Yet though capital has returned, the world has changed, as it had after the Great Depression. Much of Japan and her imitators are stuck with a late-Fordist economy, dominated by industries suffering from overcapacity. Other parts of the regional economy, notably in Taiwan, have been moving into hi-tech and IT. The diaspora as a whole has been split between losers, dragged down by speculation, debt and crony ties to discredited regimes, and winners who, with entrepreneurial diversification, have restructured and moved into software and internet services. East Asia will once again become the engine of growth but moving it in a different direction and with a different dynamic.

'The Three Faces of Capitalism and the Asian Crisis', by Constance Lever-Tracy, takes a globalising world, characterised by increasing flows between multiple and diverse centres, as the context for the Asian crisis. It seeks to identify problems of articulation between different ways of integrating capitalist operations, conceptualised in a tripartite schema of hierarchical plan, anonymous market and horizontal networks. Each of the three, it is argued, has different requirements for effective functioning, different strengths and weaknesses and they are all likely to degenerate or collapse in different ways. All three, in changing proportions, contributed to Asia's 'economic miracle', but although there is everywhere co-existence and compromise between them, the synergy is unstable. The chapter argues that a sudden change in the balance, caused by rapidly increasing flows of hot money, produced a disjuncture at the interface and an escalating crisis. A degree of insulation at this perilous interface including, on the one hand, restriction or avoidance of short term foreign borrowing and, on the other, much stricter requirements for transparency where it does occur, seem to be a necessary minimum to attenuate future crises.

Henry Yeung, 'Managing Crisis in a Globalising Era: The Case of Chinese Business Firms from Singapore', looks at firms that were able to ride the crisis, including a detailed case study of the Hong Leong group. Such firms had already learned before the crisis, how to adjust to the mismatches at the interface of the global economy. The chapter examines the role of transnational entrepreneurship in the management of crisis tendencies. It argues that some of these firms have been highly proactive in managing the negative impact of the Asian economic crisis. In particular, they engaged in pre-emptive strategies of pursuing geographical diversification, tapping into global capital markets and using non-equity investments. These strategies enabled them not only to globalise into foreign markets before the onset of the crisis, but also to continue to grow when it struck.

In Chapter 6 David Ip looks at 'Networks and Strategies in Taiwanese Business'. The Taiwanese economy as a whole was less severely affected by the crisis than either Hong Kong or Southeast Asia, but Ip shows that the big picture masks a great diversity of effects, with bankrupcies for some and expansion and upgrading for others. The personalised networks of Taiwanese firms are shown to be janus faced, opening up disastrous pitfalls in some cases and providing strength for recovery and new beginnings in others. For small firms in particular, the expansion, during the crisis, of integrated production networks seems to have become an effective strategy, not only for survival and growth but also for self transformation in the new world of globalisation. Their ability to replicate a well-tested system of integrated production network in various regions in mainland China was crucial in accounting for the rapid rise to prominence and the confidence many Taiwanese businesses displayed on the international stage, particularly in proclaiming their ambitions to make a giant leap from sunset industries to sunrise industries, and into their own design and brand names, and to enter alliances on an equal basis with Western firms.

Cen Huang reports on detailed empirical studies of the fortunes of diaspora ventures in China, and of the wide range of responses by government at various levels. Here too there was considerable variability, with some regions and firms suffering declining investment and collapsing exports (especially those with markets in East and Southeast Asia) while others benefited from cheaper imports of material and equipment and were able to redirect their sales. Few, however, indicated any desire to switch their base of operation out of China. Many believed that the crisis was a temporary setback, and all had their eyes on expanding into China's domestic market at the end of it. Nevertheless, many felt that a diversification of markets was necessary in the future, in order to strengthen and consolidate their business. It was clear too that many of the same local governments who had benefited from the profits of the diaspora Chinese investors, were now offering to remit charges and taxes and to remove red tape and other restrictions, to avoid killing the golden goose. The effect of the crisis was to drive some out, but to push those able to react to globalise their markets and to upgrade their skills and products and to move into high tech areas.

In Chapter 8, 'Responses to the Crisis: Network Production, Diversification and Transnationalism in Chinese Businesses', David Ip turns his attention to small and medium enterprises, the heartland of Chinese diaspora capitalism. He provides a work in progress report on the first 25 (out of a planned 120) in depth interviews with Chinese business owners in Hong Kong, Taiwan and Australia and concludes provisionally that there have been winners as well as losers and that survival has been widely achieved through a variety of strategies. These have included the extension of credit within trust based personal networks as well as diversification into new opportunities and new regional and global markets and the upgrading of technology and managerial effectiveness. In Taiwan there was an impetus to further development of integrated production networks. Many of the distinctive characteristics defining diaspora Chinese business have continued to be relevant during the Asian crisis. Despite the differential impact of the crisis on firms of different sizes, kinds of management and activities, many respondent owners still preferred to maintain full autonomy and control. Networks have remained for them both morally honored and significant as a mechanism for minimising debts and vulnerability. Few reported abuse of the trust and goodwill associated with such networks. Moreover, diversification and transnationalism have not been sources of weakness and unsustainable risks. Rather, in the current climate of gradual, albeit slow recovery, many still followed such strategies, while exercising more cautious and rational productivism than before and seeking to avoid speculative and risky operations.

Noel Tracy's Chapter 9, 'Weathering the Storm: Structural Changes in the Chinese Diaspora Economy', analyses the impact of the crisis on large Chinese business groups. He concludes that despite the severity of the crisis most of these groups have survived even if some are substantially reduced in size and in the scope of their activities. He notes that the Southeast Asian Chinese, particularly those in Indonesia, Malaysia and Thailand, have been hit hard and have declined in importance in terms of the overall reach and size of the Chinese business area. He also points to the large number of new start-ups that the opportunities and threats presented by the crisis have produced, indicating a major regeneration process at work. His most important conclusion, however, is that the crisis has enabled high-tech and computer-based industries to advance at the expense of more traditional sectors like property and finance and has made of the former the new key sectors of the Chinese diaspora economy. This transformation has emerged most rapidly in Taiwan and has propelled Taiwan forward as the industrial core of the whole Chinese business area. He notes, however, that these developments in Taiwan have soon been followed by similar initiatives in Hong Kong and Singapore.

The final chapter, titled with some exaggeration 'The Irrelevance of Japan', picks up the theme with which we started this chapter, of the diversionary effect of focussing on the wrong part of the picture. It took Western observers a long time to accept that Asian players could be in for the big stakes. Having finally accepted Japan as one of the economic super powers, however, they seemed to reject any further admissions. The myth of imminent Japanese regional hegemony has interfered with a serious confrontation with the reality of Chinese diaspora transnational entrepreneurship. The chapter seeks polemically, but with facts and figures, to debunk the idea that the rise of East Asia was initiated by a Japanese lead goose, that recent development and integration of the region has been managed by Japanese companies and that only a Japanese recovery could save the region from the crisis. As Japan continues to wallow in the doldrums, the rest of the region is moving on in new directions.

Notes

1. There are continually new examples of further upward revisions of growth estimates.
2. For further material on the overseas Chinese and their history see Wang (1991) and in more popular style Pan (1990).
3. For general discussions on the importance of networks, associations and *guanxi* see Hwang (1990), Redding (1990, pp. 176, 215). For more localised accounts see for

example Lever-Tracy et al (1991) on Chinese in Australia, Omohundro (1981) on the Philippines, Light (1972) on the USA, Nonini (1983) and Gosling (1983) on Malaysia and Thailand, Barton (1983) on Vietnam, Suehiro (1989, p. 121) on Thailand, Greenhalgh (1984 and 1988, p. 236) on Taiwan, Chan and Chiang (1994) on Singapore, Mackie (1989, p. 99; 1992) and Limlingan (1986) on Southeast Asia and Hamilton et al. (1990) on East Asia. In Hong Kong, where immigration has been massive and recent and the rule of law has been reliable, if remote, associations seem to play a lesser role, and networks are more often based on individual construction than on community guarantee (Sit and Wong 1989, p. 237).

[4] Mackie (1989, p. 101) comments that even the most prominent *cukongs* in Indonesia exerted indirect and personal influence, not formal power on behalf of any general business group, that many of the most successful had no such ties and that some prospered even more after losing the ties. Suehiro argues that the growth of the Bangkok Bank was not primarily due to its powerful military patron, noting that the bank continued to expand after the collapse of their patron's government (1989, p. 110).

[5] For example see Chan and Chiang, 1994; Cushman and Wang (eds.), 1988; Hamilton (ed.), 1991; Jesudason, 1989; Kotkin, 1993; Lever-Tracy et al, 1996; Lim and Gosling (eds.), 1983; Limlingan, 1986; Mackie, 1992; McVey (ed.), 1992; Omohundro, 1981; Orru, Biggart and Hamilton, 1997; Redding, 1990; Tam, 1990; Wang, 1991; Wong, 1988; Wu and Wu, 1980; Yamaguchi, 1993.

References

Barton, C. (1983), 'Trust and Credit: Some Observations Regarding Business Strategies of Overseas Chinese Traders in South Vietnam' in Lim and Gosling (eds).

Bello, W. and Rosenfeld, S. (1990), *Dragons in Distress: Asia's Miracle Economies in Crisis*, IFDP, San Francisco.

Chan, K.B. and Chiang, C.S.N. (1994), *Stepping Out: The Making of Chinese Entrepreneurs*, Prentice Hall, Singapore.

Chirot, D. and Reid, A. (1997), *Essential Outsiders: Chinese and Jews in the Modern Transformation of Southeast Asia and Central Europe*, University of Washington Press, Seattle.

Chiu, P.Y.W. (1988), *The Economy of Hong Kong*, Enterprise Publishing, Hong Kong.

Clegg, S. and Redding, S.G. (eds) (1990), *Capitalism in Contrasting Cultures*, Walter de Gruyter, Berlin.

Cushman, J. and Wang, G.W. (eds) (1988), *Changing Identities of the Southeast Asian Chinese Since World War II*, Hong Kong University Press, Hong Kong.

FEER (Far Eastern Economic Review), various dates.

Freedman, M. (1979), *The Study of Chinese Society*, Stanford University Press, Stanford.

Gerschenkron, A. (1962), *Economic Backwardness in Historical Perspective*, Harvard University Press, Cambridge.

Giddens, A. (1990), *The Consequences of Modernity*, Stanford University Press, Stanford.

Gosling, P.L.A. (1983), 'Chinese Crop Dealers in Malaysia and Thailand: the Myth of the Merciless Monopsonistic Middleman' in Lim and Gosling (eds).

Greenhalgh, S. (1984), 'Networks and Their Nodes: Urban Society on Taiwan', *China Quarterly*, 99 (September).

Greenhalgh, S. (1988), 'Families and Networks in Taiwan's Economic Development' in Winckler, E. and Greenhalgh, S. (eds), *Contending Approaches to the Political Economy of Taiwan*, An East Gate Book, Armonk.

Hamilton, G. (ed.) (1991), *Business Networks and Economic Development in East and Southeast Asia*, Centre for Asian Studies, University of Hong Kong, Hong Kong.

Hamilton, G. (1991), 'The Organisational Foundations of Western and Chinese Commerce: A Historical and Comparative Analysis' in Hamilton (ed).

Hamilton, G. (1998), 'Asian Business Networks in Transition: What Alan Greenspan does not Understand about the Asian Financial Crisis', Keynote Speech to *Workshop on Asian Business Networks*, National University of Singapore, March.

Hamilton, G., Zeile, W. and Kim, W.J. (1990), 'The Network Structures of East Asian Economies' in Clegg and Redding (eds).

Ho, R. (1997), *Diaspora Chinese Capitalism: a Chop Suey of Rational and Speculative Ventures*, B.A. Honours Thesis, Sociology, Flinders University of South Australia, Adelaide.

Hofheinz, R.J. and Calder, K.E. (1982), *The East Asia Edge*, Basic Books, New York.

Hwang, K.K. (1987), 'Face and Favor: The Chinese Power Game', *American Journal of Sociology*, 92/4 (January).

Jesudason, J.V. (1989), *Ethnicity, the Economy and the State, Chinese Business and Multinationals In Malaysia*, Oxford University Press, Singapore.

Kano, H. (1989), 'Indonesian Business Groups and Their Leaders', *East Asian Cultural Studies*, 28.

Kao, C.S. (1991), 'Personal Trust in the Large Businesses in Taiwan: A Traditional Foundation for Contemporary Economic Activities', in Hamilton (ed.).

Kotkin, J. (1993), *Tribes: How Race Religion and Identity Determine Success in the New Global Economy*, Random House, New York.

Krugman, P. (1998), 'Saving Asia', *Time,* 7th September.

Lam, D., Paltiel, J. and Shannon, J. (1994), 'The Confucian Entrepreneur? Chinese Culture, Industrial Organisation, and Intellectual Property Piracy in Taiwan', *Asian Affairs*, 20/4 (Winter).

Lever-Tracy, C., Ip, D., Kitay, J., Phillips, I. and Tracy, N. (1991), *Asian Entrepreneurs in Australia: Ethnic Small Business in the Chinese and Indian Communities of Brisbane and Sydney*, OMA/AGPS, Canberra.

Lever-Tracy, C., Ip, D. and Tracy, N. (1996), *The Chinese Diaspora and Mainland China: An Emerging Economic Synergy*, Macmillan, Houndmills.

Light, I. (1972), *Ethnic Enterprise in America: Business and Welfare Among Chinese, Japanese and Blacks*, University of California Press, Berkeley.

Lim, L.Y.C. and Gosling, P.L.A. (eds) (1983), *The Chinese in Southeast Asia, Volume 1 Ethnicity and Economic Activity*, Maruzen Asia, Singapore.

Limlingan, V.S. (1986), *The Overseas Chinese in ASEAN: Business Strategies and Management Practices*, Vita Development Corporation, Manila.

Mackie, J. (1988), 'Changing Economic Roles and Ethnic Identities of the Southeast Asian Chinese: A Comparison of Indonesia and Thailand' in Cushman and Wang (eds).

Mackie, J. (1989), 'Chinese Businessmen and the Rise of Southeast Asian Capitalism', *Solidarity*, 123 (July/September).

Mackie, J. (1992), 'Overseas Chinese Entreprenurship', *Asian-Pacific Economic Literature*, 6/1 (May).

McVey, R. (ed.) (1992), *Southeast Asian Capitalists*, SEAP, Cornell University, Ithaca.

Nonini, D.M. (1983), 'The Chinese Truck Transport Industry of a Peninsular Malaysia Market Town', in Lim and Gosling (eds.).

Omohundro, J. T. (1981), *Chinese Merchant Families in Iloilo. Commerce and Kin in a Central Philippine City*, Ateneo de Manila University Press, Quezon City.

Orru, M., Biggart. W. and Hamilton, G. (1997), *The Economic Organisation of East Asian Capitalism*, Sage Publications, Thousand Oaks.

Oxfeld, E. (1993), *Blood, Sweat and Mahjong: Family and Enterprise in an Overseas Chinese Community*, Cornell University Press, Ithaca.

Pan, L. (1990), *Sons of the Yellow Emperor*, Secker and Warburg, London.

Redding, S.G. (1990), *The Spirit of Chinese Capitalism*, Walter de Gruyter, Berlin.

Robertson, R. (1992), *Globalisation: Social Theory and Global Culture*, Sage, London.

Sachs, J. (1997), 'The Wrong Medicine for Asia' reproduced from New York Times 1997 //www.stern.nyu.edu/~nroubini/AsiaHomepage. html. Asiasach.doc

Sangren, P. S. (1984), 'Traditional Chinese Corporations: Beyond Kinship', *Journal of Asian Studies*, XLIII/ 3 (May).

Sato, Y. (1993), 'The Salim Group in Indonesia: The Development and Behaviour of the Largest Conglomerate in Southeast Asia', *The Developing Economies* XXXI/4 .

Sit, V.F.S. and Wong, S.L. (1989), *Small and Medium Industries in an Export-Oriented Economy. The Case of Hong Kong*, University of Hong Kong, Hong Kong.

Suehiro, A. (1989), 'Bangkok Bank: Management Reforms of a Thai Commercial Bank', *East Asian Cultural Studies*, 28.

Suryadinata, L. (1988), 'Chinese Economic Elites in Indonesia: a Preliminary Study', in Cushman and Wang (eds.).

Tam, S. (1990), 'Centrifugal Versus Centripetal Growth Processes: Contrasting Ideal Types for Conceptualising the Development Patterns of Chinese and Japanese Firms', in Clegg and Redding (eds.).

Wang, G.W. (1988), *China and the Chinese Overseas*, Times Academic Press, Singapore.

Wong, S.L. (1988), 'The Applicability of Asian Family Values to Other Sociocultural Settings', in Berger, P.L. and Hsiao, M.H.H (eds), *In Search of an East Asian Development Model*, Transaction Books, New Brunswick.

Wu, Y.L. and Wu, C.H. (1980), *Economic Development in Southeast Asia: the Chinese Dimension*, Hoover Institution Press, Stanford.

Yamaguchi, M. (1993), 'The Emerging Chinese Business Sphere" *Nomura Asian Perspectives*, 11/2 (July).

2   Chinese Diaspora Capitalism on the Eve of the Crisis

CONSTANCE LEVER-TRACY

Introduction

By the mid 1990s it was becoming increasingly hard to ignore or deny the growing economic strength, confidence and transnational interlinkages of Chinese diaspora capitalism. Loosely woven, flexibly shifting networks of family businesses, without a coordinating centre or bureaucratic hierarchy, were clearly playing a central role in the dramatic development of new *little tiger economies* and in the industrialisation of Southeast Asia. It was, above all, diaspora Chinese businesses who were, by far, the major investors in Mainland China, acting as partners and channels for modern technology and production methods for her exploding rural industries and as their bridge to world markets - in sharp contrast with the ineffectiveness of the multinationals either in China or in the ex Soviet empire.

News magazines and business journals[1], started featuring their achievements as did popular books with such titles as *Sons of the Yellow Emperor* (Pan, 1990), *Tribes, How Race, Religion and Identity Determine Success in the New Global Economy* (Kotkin, 1993), *Lords of the Rim: The Invisible Empire of the Overseas Chinese* (Seagrave, 1995) and *The Bamboo Network: How Expatriate Chinese Entrepreneurs Are Creating a New Superpower in Asia* (Weidenbaum and Hughes, 1996). Chinese business people themselves, who had had a tendency to avoid a high profile, began asserting themselves more visibly, holding *World Conventions of Chinese Entrepreneurs,* biannually since 1991, with the publicised presence of Chinese tycoons from around the world[2].

Academic literature had, however, surprisingly little to say about these developments. Studies of the transnational connections and activities of what we have diaspora Chinese capitalism, have long been bedevilled by the area boundaries of academic disciplines and their expertise and by accumulation of country studies, or of research that stopped at the borders of Southeast or of East Asia or of China, or which focussed only on ethnic or immigrant business in Western countries, but little that paid attention to the interlinkages. Meanwhile those studied, small owners as well as billionaire tycoons, had increasingly been moving their products and investments, their family members and themselves back and forth across these boundaries, facilitated by complex, historically constructed transnational networks.

Perhaps more important than the political boundaries and those of specialisms and linguisic expertise were the theoretical blinkers. Those interested in explaining the unprecedented developmental *miracles* of the region had their attention focussed on the repetitious debate between *strong states* and *free markets*, forgetting that while these might structure the pattern of opportunities, there still had to be active agents, able and willing to respond to them.

Those who did focus on the overseas Chinese were, with some notable exceptions (including Wu and Wu, 1980; Limlingan 1986; Wong, 1988), trapped by theories that confidently predicted the impossibility of just those developments that were in fact occuring. Such theories can be schematically seen as of two kinds (although they also overlap). The first are those that assumed the weakness, limited horizons and restricted roles of their small family businesses, in a world increasingly dominated by giant corporations. Insuperable obstacles to growth were said to be posed by their lack of state support or access to capital. The most recent, somewhat eccentric, version was by Francis Fukuyama (1995). He argued that familistic societies, particularly those of the Chinese, had a long tradition of mistrust of non kin. The refusal to trust professional managers and the subdivision that occured with partible inheritance meant that few firms could grow large or remain so over several generations. The low level of general trust also prevented these small companies from networking to compensate for their lack of economies of scale, thus they offered no serious competition to 'high trust' societies with their large and long lived firms.

The second approach goes back to Max Weber's contrast between the impersonality and the market based or bureaucratic rationality of modern capitalism and the personalistic or ascriptive nature of traditional social forms (both small and large). Chinese traders were an archetype of the traditional capitalist middlemen of classical sociological theory. They might have a temporary role, in transitional contexts, filling a narrow niche as intermediaries between indigenous producers and colonial states or modern capitalist corporations, until the development of modern institutions makes them redundant. Linda Lim (1983a and b, pp. 6, 245) argued that they would have to abandon their distinctive Chinese characteristics, if they were to survive in technologically based industries oriented to the world market.

There have been two mutually contradictory and only partially satisfactory attempts to revise these presumptions in light of an evolving reality. The first, particularly the work of Redding (1990; 1991), and of Tam (1990), focused in particular on the recent achievements of smaller firms in Hong Kong and Taiwan. This approach saw Chinese businesses as constituting a dynamic system of 'entrepreneurial familism', of small but networked firms which were in fact well suited to operating in the contemporary world[3]. Both Redding and Tam, however, denied that such firms had either the ability or the will to grow, suggesting that larger Chinese firms might be seen as not only unusual but aberrant. Chinese businesses were well suited to subcontracted manufacturing of a narrow product range and had no capacity or ambition to enter the world of design, marketing or brand names (Redding,1990, p. 229).

The second approach (well represented in McVey, ed., 1992), which looked exclusively at Southeast Asia, saw a new developmental dynamism in a constructive alliance of local big business and the state. While some of these writers were more willing than others to acknowledge the ethnic composition of these big conglomerates, there was little place in the model for any discussion of the contribution or changing nature of the main body of Chinese firms, which were of small and medium size and which lacked the particularistic state ties of the big ones. Indeed they were explicitly excluded. McVey explained that the writers she had collected in her book restricted themselves to a focus on big business because, as she asserted somewhat *a priori*, '..it is at this level rather than that of petty and middling entrepreneurs that Southeast Asian capitalist development has been most marked in the last two decades' (1992, p. 9). Similarly Mackie, confronting the apparently limited explanatory power, for recent developments, of the large prior literature on the small businesses of the overseas Chinese, suggested one feels in some desperation, that both the literature and its subject matter might be of little relevance in looking at the recent emergence of the large conglomerates. These latter day Chinese should perhaps be regarded as 'a quite different kind of Chinese from their predecessors' (Mackie, 1992, p. 59).

With the onset of the Asian crisis this revision, which saw such tycoons in a progressive alliance with Asian states, has given ground. In its place comes a revival of denunciations of '*ersatz* capitalism' and of traditionalist, rent seeking 'crony capitalism', with its personalistic ties to the politically powerful and renewed calls for their replacement by a Western model of professional and bureaucratic impersonality.

This chapter will first describe the real and substantial achievements of diaspora Chinese business, on the eve of the crisis. It seems clear that these successes were not a consequence of the simple replacement of distinctively Chinese by Western business culture and forms of organisation. Nor were they restricted either to small firms, subcontracting contentedly in narrow niches, or to a group of unrepresentative state sponsored tycoons. In the final section we will return briefly to the theories sketched above, and suggest reasons for their inadequacy. While discussion of such issues seemed shelved by the Asian crisis, the recent recovery puts it again on the agenda.

The Successes

The Chinese Business Sphere

It was in Japan that awareness of this rising economic power of Chinese business was most salient. Yamaguchi proposed the term 'Chinese Business Sphere', to 'refer to the region where Chinese were actively doing business using contact networks' (1993, p. 3). Its core he constituted by combining China, Taiwan, Hong Kong, Macao and the ASEAN countries[4]. To this we could add Vietnam. He noted that even in those countries within this Chinese Business Sphere where ethnic Chinese constituted a small proportion of the population, they were responsible for a large or dominant proportion of private capitalist business activity. This economy, he warned Japan, was 'emerging from dependence on developed economies such as Japan, the US and Europe into an independent growth pattern' and it had the potential to lead the world economy in the coming century (p. 19).

Much evidence could lend support to such a claim. Over three decades in East Asia, and two in Southeast Asia, economies had sustained rates of economic growth of between 5% and 8% each year. In China under reform, growth rates in many years exceeded 10%. Per capita income levels increased fivefold in Thailand, and fourfold in Malaysia, and in Hong Kong and Singapore per capita purchasing power came to exceed that in Japan.

Table 2.1 Per Capita GNP on the Eve of the Crisis (US$)
(GDP per capita at Parity Purchasing Power, in brackets)

Japan	33,800 (23,840)	Malaysia	4,287	(11,700)
Singapore	31,900 (28,780)	Thailand	2,450	(6,940)
USA	29,950 (30,025)	Philippines	1,203	(3,565)
Hong Kong	26,400 (25,500)	Indonesia	998	(3,790)
Australia	20,020 (21,470)	China	738	(3,650)
Taiwan	13,303 (16,610)	India	387	(1,680)
South Korea	9,511 (13,990)			

Source: Adapted from *Asiaweek*, 21 August, 1998, p. 67.

While the benefits were far from evenly spread, some measures of inequality, in Taiwan for example, and between rural and urban dwellers in China, fell sharply[5], and throughout the region the impact on the lives of ordinary people was felt in substantial reductions in infant mortality, illiteracy and the proportions living in absolute poverty and in rising life expectancy and levels of consumption and of secondary education[6]. Such sustained growth in what had been impoverished countries, was unprecedented elsewhere in the world, in a period when South Asia had stagnated[7] and Latin America and Africa had gone backwards. The relative position in the world of the countries of East and Southeast Asia changed dramatically[8].

By the early 1990s Yamaguchi estimated his Chinese Business Sphere to be the world's fourth largest economic power bloc, after the US, Europe and Japan, with an expected average annual growth rate in the 1990s of 7.9% compared with a world average of 2.8%. Such forecasts were not too far off the mark until the crisis struck in mid 1997 (1993, p. 19).

The Chinese Business Sphere, Yamaguchi noted, held 24% of the world's foreign exchange reserves in 1991 (up from 20 % in 1988) and had taken the leading role from Japan on the world scene of mergers and acquisitions (1993, p. 5). By mid 1997, Taiwan, Singapore and Hong Kong held between them $296 billion in international reserves, and China held another $123 billion, compared with Japan's figure of $221 billion (*FEER*, 9th October, 1997, pp. 86/87).

In 1993, for the first time, the combined imports and exports of China, Hong Kong and Taiwan surpassed the total trade of Japan (GATT secretariat, cited in Kojima, 1994, p. 18). China's ranking as a global exporter had moved up faster than any other country in the world. In 1978 China ranked 32nd in the world as an international trading economy, in

Table 2.2 Economic Growth in East and Southeast Asia Before the Crisis

	1995	1996	1997
East Asia			
China	9.9	9.6	8.8
Hong Kong	5.0	4.6	5.3
Taiwan	6.6	5.7	6.8
Southeast Asia			
Indonesia	7.2	8.0	4.9
Malaysia	9.2	8.6	7.5
Philippines	5.6	5.8	5.2
Singapore	8.3	6.9	8.0
Thailand	8.6	5.6	-1.3
Japan	1.0	3.8	-0.7

N.B. Thailand was the country earliest affected by the crisis.

Source: Adapted from *Far Eastern Economic Review*, 26 December, 1996, pp. 132/3; 24 December, 1998, pp. 58/9; 14 October, 1999, pp. 72/3.

1992 she ranked 11th (Yan, 1995, p. 10). In 1994 Kojima estimated that 'the combined economy of the three Chinas', China, Hong Kong and Taiwan, had 'replaced Japan as the major player in the Asian economy' (p. 18).

Such information about the countries in which ethnic Chinese businesses are important is only indicative for it covers both too much and too little. There are many other players in these countries, including not only indigenous businesses and the national state but also other migratory ethnic groups and other foreign capital, particularly American, European and Japanese multinationals. On the other hand Chinese diaspora networks have crucial nodes in places like Vancouver, Silicon Valley and Sydney, and some presence in most countries around the world.

Economic Strength of the Chinese Diaspora

In 1992 *The Economist* estimated the liquid assets of the somewhat over 50 million ethnic Chinese living outside the Mainland as between $1.5 and $2 trillion, half as much as the total value of Japan's bank deposits, a gap that must have narrowed substantially over the next five years. However, such estimates of the economic power of Chinese capitalists, operating within this region, are of necessity scarce, patchy and speculative and bear no consistent relationship either to the size of the Chinese population within a country or to the country's economic indicators. In addition information is much easier to find on the changing fortunes of high profile tycoons and on the larger, publicly listed companies than on the majority of small and medium family businesses. Yet these latter are often not only of more cumulative weight, but also the cutting edge of economic dynamism and transnational orientation.

Some of the clearest data come from Malaysia, where the government policies of seeking to develop an indigenous Malay capitalism have led to the systematic collection of information on the ethnicity of capital ownership. In 1971, 62% of the private corporate economy was foreign owned, 34% owned by local ethnic Chinese and Indians and only 4% owned by Malay Bumiputras. By 1985 a policy of favouring Bumiputras had raised their share to 18%, while foreign ownership had fallen to 25%. The share of Chinese and Indians, which it was intended should remain unchanged, had risen to 57% and at least 55% of the non foreign corporate economy was in Chinese hands (Redding, 1990, pp. 30, 31). In Singapore on the other hand, despite being a largely Chinese city, the heights of the economy remained dominated by Western and Japanese multinationals, although locally owned companies had been gaining ground.

The shifting power balance, between British and Chinese, at the summit of the Hong Kong economy, is documented by Gilbert Wong (1991), who analyses the changing pattern of interlocking directorates among the largest listed public companies and banks between 1976 and 1986. He traces the rise of large ethnic Chinese business groups and how they displaced the British, by going into manufacturing and property, where they were able to raise the money to take them over. In 1976 'the major business groups were made up of and controlled by non-Chinese business families or organisations' (p136). The Chinese made a 'triumphal entry' in 1981, and by 1986 British hongs had been largely replaced by Chinese in the business structure, with the Li Ka-shing group at the top. Foreign ownership in manufacturing was always quite limited, probably never employing over 10% and touching no more than 1% of enterprises (Chiu, 1988, p. 39). Both foreign and local British capital was concentrated in financial and trading sectors with industrialisation initiated primarily by local Chinese entrepreneurs. Even subcontracting to overseas customers and trading companies affected under 5% of small and medium manufacturers surveyed in 1987 (Sit and Wong, 1989).

In Taiwan foreign companies have never played a central role in the growth of the economy or in its export orientation or overseas investment. Even in 1964 domestic savings accounted for two thirds of capital formation, and 95% by 1969 (Evans 1987, p. 206). On the other hand much data, especially on the trade and foreign investment activities of Taiwanese companies, is unreliable or unavailable either because of their exclusion from international bodies (and their statistical compilations) or because of their desire to evade their own government's regulations.

As table 2.3 shows, even in those Asian countries where the diaspora Chinese are quite small minorities, local ethnic Chinese predominate in the ownership of non foreign public companies[9]. It was estimated that, already by the mid-1980s, local Chinese capital was a more important part of Southeast Asian capital than was that from all foreign sources.

Table 2.3 Ethnic Chinese in ASEAN

Ethnic Chinese In:	Population 1991 (bn.)	% of Pop.	Share of listed equity
Indonesia	5.81	3.5%	73%
Malaysia	5.33	29%	61%
Philippines	1.20	2%	50%
Singapore	2.14	77%	81%
Thailand	5.57	10%	81%

Source: *Asiaweek*, 20 October, 1993, p. 58, reproduced from Nomura Research Institute.

In 1989 Jamie Mackie reported identifying 'at least 40-50' large scale Chinese conglomerates in Southeast Asia (p. 97). Three years later he had increased this estimate to 'over 70' (1992, pp. 161, or 183/4). It is not clear whether he had found more of them or whether their number had increased in this short interval. The largest were to be found throughout the region (see Table 2.4).

Table 2.4   The Largest Ethnic Chinese Owned Companies (including state firms)

	China	Taiwan	Hong Kong	Southeast Asia
Largest 50	22	15	6	7
Largest 500	185	90	125	100

Source:Adapted from *Asiaweek* (Chinese Language Edition), 24 October, 1993.

In Indonesia in the 1980s, of the 146 business groups whose ethnic ownership could be identified, half were majority Chinese owned. This proportion increased with the size of the group, to over two-thirds of those with over 30 companies in the group (Kano, 1989, p. 150). In Thailand four of the five largest banks and in Singapore three of the four largest are Chinese owned (*The Economist*, 18 July, 1992). In 1983 the *Far Eastern Economic Review* had proclaimed the 'Birth of a Multinational' with its story on the companies associated with Liem Sioe Leong, a group of whose activities they said 'few people had taken any notice' a year earlier. In that year there were between 40 and 55 companies in the Liem family's group, employing 30 000 people. By the early 1990s the Liem empire in Indonesia included close to 400 companies and employed at least 140 000 people. His group was estimated to produce at least 5% of Indonesia's entire GDP. *Fortune International* (28 June, 1993) ranked the Liem family as 63rd richest in the world. Chinese family fortunes are made up of multiple small firms. Over this decade of rapid growth the average size of the Liem companies in Indonesia fell from an average of 500-750 employees each at the earlier date to an average of 350 each in the early 1990s[10].

The Liem case is not untypical. The *Forbes Magazine* list of billionaires in 1994 showed that a fifth of the top 25 personal fortunes in the world (with over $5 billion each) were held by ethnic Chinese **families**, as were 15% of the 143 with over $2 billion and 10% of the 350 with over $1 billion. However, there is no such prominence when we look at the size of Chinese owned **companies**. *International Business Week's* list of the top 1000 world companies in 1993 (12 July, 1993, pp. 41, 44-81) includes only

28 Chinese owned companies, 2.8% (none of them from the Liem stable). There were only two in the top 200 companies and none in the top 100. Of all the companies belonging to the top five Chinese families (those who ranked in the wealthiest 25 families in the world), only one ranked higher than 200th. In Fortune's list of the largest 'Global 500' companies for 1996 there were 162 from the United States and 126 from Japan. There was only one Chinese owned company (Chinese Petroleum, from Taiwan) ranking 458th (4 August, 1997).

The irruption into the Western public eye of billionaire families and regional conglomerates is only the most obvious, and not necessarily the most important development. Others have argued that in the most successful economies of the Chinese business sphere, Hong Kong and Taiwan, it has been the small and medium family firms that have predominated and even increased their relative weight and that it is these that have been the spearhead of export oriented growth.

In Hong Kong two thirds of the small and medium firms surveyed in 1987 were engaged in exporting and although they frequently operated through a growing multitude of import-export houses these also were generally very small indeed, with an average size of six employees (Sit and Wong, 1989, p. 149; Tam, 1990, p. 169; SCMP, 2.7.94. The average size of manufacturing firms had fallen and the small and medium firms' proportion of gross output had increased between 1972 and 1985 and their number and relative weight continued to grow thereafter as the economy expanded. In 1989 61% of workers were employed in establishments of under 100 workers, producing 53% of value added, but only 14% in those with over 500 workers producing only 18% of value added (Hong Kong Digest of Statistics, 1991; Sit and Wong, 1989, p. 29).

In Taiwan in the early 1960s came a change, from a policy oriented to heavy industry, for armaments, and import replacement to export oriented industrialisation. This led to a shift from large, mainly mainlander and government ownership, to the growth of local small and medium firms. By the latter part of the 1980s only 40 factories employed over 1000 workers and only 1500 over 100. Twenty eight thousand had under 50 employees of whom 20,000 had fewer than ten. Over 90% of businesses were classified as small and medium family firms (Hsiao and Hsiao, 1989; Rabushka, 1987, p. 164; Lin, 1991)[11]. Even in Singapore the proportion of small firms, after declining until the early 1970s, began to rise again thereafter and Lee Kwan Yew sought to reorient public policy towards encouraging them (Chew, 1988).

Upgraded Technology and Increased Autonomy

Borrus (1994) and Hobday (1995) have both described how Americans, in the 1990s, regained their global lead in electronics from the Japanese, in large part by constructing a mutually supportive alliance with the growing drive, skills and innovative capacity of East Asian engineers, managers and firms.

Hobday describes, with a wealth of empirical detail, the 'bold, deliberate and creative strategies' (p. 201) initiated and developed by the 'latecomer firms' in South Korea, Taiwan, Hong Kong and Singapore, which enabled them to seek out and profit from such alliances. This allowed 'masses of latecomers to subcontract their way up the technology ladder, develop new industries, build innovative capabilities and catch up with industrialised countries' (p. 8). By the 1990s each had developed significant capabilities in at least some areas of advanced electronics. They had gained control over modern manufacturing processes, started to design new products and had created distinctive new organisational forms for competing on the international stage. Where necessary they were buying companies to get access to guarded technology, and they were beginning to pressurise Japan from below in key areas such as computers and semiconductors.

The process was not always easy, and only a few such as Taiwan's ACER were, at the onset of the crisis, reaching the final stage by establishing their own known brand name and developing new products[12] a capacity that still eluded most East Asian firms. The NIEs were still significantly dependent for capital goods and some key high tech components on purchases from Western and Japanese suppliers. Nonetheless, Hobday considers their progress so far to be 'almost unparalelled in economic history' (p. 94 n.21) and that many were now approaching the innovation frontier.

Asiaweek (5 January, 1994, p. 45) reported examples in 1994, including a Hong Kong 'home grown electronics giant', which had been buying subsidiaries of ailing companies in UK, Thailand, Hong Kong and Italy with a strategy of going for high value niches and custom design, planning to sell whole multi-media systems within five years. Other examples were Champion Technology (that in 1993 bought the British company that first manufactured electronic pagers) and Multitone, with plants UK and Malaysia (*FEER*, 13 April, 1994, p. 53).

Such stories are not restricted to electronics. Data on over 4000 garment factories and agents in Hong Kong, which owned, operated or contracted to over 18,000 garment factories in south-eastern China, found that the most active American labels to order from them were those involving quality, up market, high fashion, rapid change products. A survey of over 400 foreign invested firms in China, producing a variety of goods found that 28% carried out all their own design and another 18% some of it (Christerson and Lever-Tracy, 1996, pp. 579/580). Hobday's most telling story is of how bicycle production, including manufacturing, design and brand names was transferred from the US to Taiwan in little more than a decade. Attempting to break a strike at home, the American company, Schwinn, a world leader, had transferred production to its subcontractors in Taiwan and China. These came to the forefront in developing new carbon frames and mountain bikes, while the parent proved unable to keep up with technological change and unsuccessful at preventing the introduction of their erstwhile subcontractors' brands into the US. In 1992 Schwinn filed for bankrupcy (1995, pp. 129-132).

Regional Linkages

In the 1980s, manufacturing, finance and markets were developing within East and Southeast Asia, with progressive integration of regional trade and investment flows. The previously discrete Chinese trading or manufacturing communities around the region now had the motive and opportunity to start diversifying and linking up[13], thus benefiting from and contributing to these developments, a process that was massively promoted by the opening up of China.

A focus on such regional trade and investment draws our attention to the growing regional ties, which were some of the most potent factors in the growth of the countries of the region and their effectiveness as players in the global economy. Insofar as these linkages followed the pathways of their transnational networks, they demonstrate the role of the Chinese in the rise of the region more clearly than does an emphasis on their contribution to or profit from separate domestic economies. While many tycoons owed their position in part to their close relationships with governments in their adopted homelands, a region-wide analysis shows that they were able to operate just as effectively through their transnational networks in places where they had no such political sponsorship. A good example here is again provided by Liem's Salim Group in Indonesia. Liem Sioe Leong was one of the principle business partners of President Suharto and a major beneficiary of that connection. By the early 1990s, however, more than a third of his group's activity was outside Indonesia (EAAU, 1995, p. 163), profits from which were to sustain them during the Indonesian crisis and after the fall of Suharto.

Such a focus on the broader picture also reduces the importance of identifying the (sometimes deliberately) murky national identity of actors whose activities are regional in scope and often channelled through the central nodes of Hong Kong and Singapore. Much of Taiwanese trade and investment abroad, especially in China, was for long concealed and directed via Hong Kong, to avoid their own government's restrictions. Southeast Asian Chinese capitalists have also often carried out their regional activities through affiliates or subsidiaries based in Hong Kong or Singapore, to avoid accusations of disloyalty. Of the top 200 companies by capitalisation on the Hong Kong Stock Exchange at the beginning of 1995 no less than 26 (13%) were owned and controlled by Southeast Asian diaspora interests[14] and it was most often through these that their activities in China and around the region were directed.

Trade

The volume of trade of a region approximating to Yamaguchi's Chinese Business Sphere (including China, HK and Taiwan and the countries of Southeast Asia - Philippines, Indonesia, Singapore, Thailand, Malaysia and Vietnam) grew slowly between 1980 and 1985 and then more than doubled by 1990. By 1996 it had increased by another 126%, over a period when total world trade increased by only 56%. Intra regional trade, between the countries of this sphere, increased even faster during the 1990s, up by nearly 160% from 1990 to 1996. All parts of the region experienced rapid and substantially accelerated growth. The exports of these countries also more than doubled between 1990 and 1996, reaching a combined figure of US$784 billion. Each individual country experienced similar trends to varying degrees (calculated from *DOTS*, various dates).

Trade between the two great Chinese entrepots of Hong Kong and Singapore (including the re-exports they channel in both directions between China and Southeast Asia) increased at the same rate as intraregional trade as a whole. Hong Kong's trade with Singapore was up from under US$6 billion in 1990 to US$15.5 billion in 1996 (*DOTS*,1997). Hong Kong became Singapore's fourth largest trading partner while the latter ranked fifth among all Hong Kong's trading partners. In his seminal article (1997), Friedrich Wu, the Head of Economic Research at Singapore's DBS Bank, notes the increasing complementarity rather than competition between the two in linking their respective hinterlands in Southern China and Southeast Asia.

Investment

Few would doubt that the most unambiguous source of international economic power is derived from direct foreign investment, which provides control over profits and management. While domestic savings have been important in the development of the countries of the Chinese Business Sphere, foreign capital has also been significant, and it is the Chinese diaspora which has been its largest provider. Although it goes outside the boundaries of this chapter, we might note that when the financial crisis struck the countries of Asia, all kinds of short term capital, in loans, traded equity or currency holdings were halted and withdrawn. Foreign direct investment (FDI) however, not only remained but continued to flow in to the countries of the region (except Indonesia whence, after ethnic riots, Chinese capital also fled) even if sometimes at a reduced rate (*FEER,* 8 April, 1999, p. 62).

In the 1990s the Chinese NICs emerged as major investors in all parts of the region, including China. Already by the late 1980s they were jointly on a par with Japan, and after 1991, as Japan substantially reduced its foreign investment they clearly overtook her. In 1992 the Nikkei Weekly wrote that the rising outflow of funds from Singapore, Hong Kong and Taiwan was replacing Japanese money as 'the main diet for the capital hungry Asia Pacific region and beyond' (30 May, p. 25).

In 1996 the four NIEs, together with the countries of Southeast Asia and mainland China, provided 14.1% of total world flows of realised FDI, most of it directed to the countries of their own region[15]. There is no reliable way of allocating this money between individual sources, since so much money (particularly from Taiwan and Southeast Asia) flowed through Hong Kong and Singapore, but there can be no doubt that a large majority of it comes from diaspora Chinese sources. In contrast, Japan that year provided only 7.0% of the world total, most of it directed outside the region (see Chapter 10).

In Taiwan exchange controls were lifted in 1987. With what were, until Japan retrenched, the largest foreign currency reserves in the world, the Taiwanese government actively encouraged investment around the region (but not in China). While the cumulative value of foreign investment outflows from Taiwan between 1959-1986 was only $272 million, over the subsequent four years its direct foreign investment was over $16 billion. By

1994 it was estimated by the Taiwanese Ministry of Economic Affairs that thousands of Taiwanese companies had poured over $25 billion into Southeast Asia. In 1993 Taiwan was the largest single investor in Vietnam and the second largest in Malaysia (*FEER*, 9 April, 1992, p. 59, 18 March, 1993, pp. 44,45, 10 June, 1993, p. 48, 16 June 1994, p. 81, 6 October 1994; Lever-Tracy and Tracy, 1993, p.18; *SWB* FEW/1321 WF/1, 23 February 1994, FEW/0319 WF/1, 9 February 1994).

Despite the strong presence of foreign multinationals in Singapore, in its regional investments Chinese capital predominated. Of the $9.3 billion invested abroad by Singapore companies between 1981 and 1991, $4.9 billion came from locally controlled companies, and only $4.4 billion from foreign invested firms. Nearly two thirds of the 2,467 companies set up overseas were by locally controlled firms and they were responsible for a large proportion of the investment within the region. Foreign owned, Singapore based firms tended rather to invest in Australia, New Zealand, the United States and Europe (*STW*, 11.12.93, p. 20).

Singapore had always invested heavily in Malaysia and also had had close ties with Hong Kong, but its investment increasingly spread around the region. In 1982 Singapore had invested $5 billion in Thailand. Eleven years later the sum was 46 times greater. By 1993 accumulated investment in Vietnam by Hong Kong ($1064 billion) and Singapore ($473 billion) placed them in rank just behind Taiwan and ahead of Australia ($415) and Japan ($414 billion). Singapore also became the largest foreign investor in Indonesia in 1993 with $679 billion (15 times more than the previous year) and Hong Kong was second with $110 (followed by UK with $114 billion and Japan with only $61 billion) (*FEER*, 29.10.92, pp. 8, 43, 3.3.94, p. 57; *SWB* FEW/0319 WF/1, 9.2.94; *STW*, 12.2.94, p.11).

The most dramatic flows of foregn investment and the ones which are most overwhelmingly attributable to the Chinese diaspora, are those which have transformed the coastal provinces of China into major competitive global exporters. China has been the largest recipient of FDI in the world in the 1990s. In 1997 half the exports from Guangdong and over 40% of the national total came from foreign invested firms (*China's Customs Statistics, Monthly*, 12, 1997).

Diaspora investment has been flowing in in significant quantity only since the Pearl River Delta was opened in 1985 but their relative contribution and the total amount increased rapidly especially after 1991. Between 1979 and 1993 there had been 167,000 foreign invested enterprises established in China, involving a total direct foreign investment of US$68.7 billion. The share of the United States in these cumulative sums was under 7% and of Japan just over 4%. Over 80% came from sources where a large majority of investors would have been ethnic Chinese: Hong Kong, Taiwan, Macao, Singapore and Thailand. Three quarters of accumulated intra regional direct investment flows to China, from 1986 to 1991, came from (or through) Hong Kong (Yamaguchi, 1993, p. 17). The inflow of 1991 doubled in 1992 and more than doubled again in 1993. The rising levels of investment continued with a further US$33.7 billion in 1994, US$35.8 billion in 1995, US$40.8 billion in 1996 and an estimated US$45.3 billion in 1997 (*WIR*, various dates).

The investment activities of high profile tycoons are sometimes highly visible. In October 1993 for example, Dhanin Chearavanont, head of the Sino Thai CP conglomerate, featured on the front cover of *The Far Eastern Economic Review* as 'The top investor showing the way into China'. By a media search between 1992-1994, we were able to collate information on major transnational direct investment projects, by the 75 largest ethnic Chinese conglomerates in Hong Kong and Macao, Taiwan and Southeast Asia. Eight out of ten of them were active beyond their own borders, 93% of those based in Hong Kong and Macao, 80% of those in Taiwan and 69% of those in Southeast Asia. The direction of their economic activity did not respect any theoretical divide between East and Southeast Asia. Of the 40 in Hong Kong, Macao and Taiwan, over a quarter were crossing borders within East Asia (to each other or to Japan) but over 40% were in Southeast Asia. Conversely, of the 35 Southeast Asians, 43% were active in one or more other Southeast Asian countries, but over half were in East Asia, 37% of them in Hong Kong. Every Southeast Asian country was receiving investment from at least three of the companies in East Asia and every one had at least two families investing in East Asia. Twenty five were also investing in North America or Europe and 11 in Australia and New Zealand. Two thirds of them had known investments in China, including 78% of those based in East Asia and 54% of those in Southeast Asia (Lever-Tracy et al, 1996, Ch. 7).

It is clear that these rapidly growing conglomerates, many larger than the regional presence of the multinationals, could not easily be portrayed as dependent or as niche bound to a *compradore* role. Their activities had become increasingly diversified, their capital and markets were increasingly found within the region and while many had links of various kinds with Western and Japanese companies, these did not involve any high level of capital or market dependency and were balanced by increasing linkages with each other[16]. The relations between these large Chinese businesses and the states where they resided had also seen a growing bargaining power on their part, a product of their own greater economic strength and of the economic linkages they had created around the region.

However, while broad data are not available, there is much evidence that the small and medium firms that predominate in the diaspora had been also operating transnationally, sometimes leading the way for the conglomerates. Our own study of small immigrant Chinese businesses in Australia, in the late 1980s, found a third of them involved in some trading and investing activities around the region.

In China small investors had been particularly important as joint venture partners of the thriving rural and suburban Township and Village Enterprises (owned most often by low level local governments). Sit and Wong's study of small and medium enterprises in Hong Kong (1989) found that, even as early as 1987, nearly a fifth of their respondents had operations in China. Hsing (1998) describes in detail how the Taiwanese shoe making industry, composed of networks of small firms, coordinated by a multitude of small export agents, transplanted themselves to China in the early 1990s. In 1993 the average foreign investment in China was only US$310,000 and in 1996 barely US$1 billion (CSY, 1997). The average foreign-invested enterprise in the southeast employed around 150 workers and in Jiangsu it was less than 100 (*Beijing Review*, March 23rd 1994, pp. 11, 14-20, 37; *South China Morning Post*, 3/4 December, 1994, pB1).

These rising intra regional flows of trade and investment have largely followed the networks of diaspora Chinese capital, with Hong Kong and Singapore as crucial intercommunicating financial nodes and entrepots (Wu, 1997) and with marketing and technology linkages extending to the West Coast of America and Silicon Valley (where many Chinese from the region work).

It was mainly in the period since the 1970s that the Chinese diaspora made the breakthrough from a concentration in trade to a prominent position also in the finance and manufacturing of the East and Southeast Asian region, from prosperity to economic wealth, and from a status as middlemen, between local producers and imperialist economic powers, to that of an autonomous local capitalist class and increasingly a global actor.

Every attempt to attribute a local range or fixed roles and narrow niches to Chinese capitalism failed, as they moved to diversify and extend their spheres of activity. Redding's belief, based largely on the Hong Kong experience at a particular time (1990, p. 229), that Chinese businesses were content with limited subcontracted manufacturing roles was soon disproved. The Thai Kanajanapas' conglomerate, for example, established chains of watch, optical and clothing shops around the region, as did Hong Kong retailers like Giordano and Toppy and Li Ka-shing's Park'n Shop. Local brand names such as Laser Computers were becoming widely known, and Hong Kong was now a design and marketing centre for goods produced in China. Limlingan's typology of Chinese managerial systems, which classified those in East Asia as *production oriented* and those in Southeast Asia (with the exception of *mixed* Singapore) as *commodity oriented* (1986, p. 155) had also lost its applicability within a few years. Of the 100 largest Chinese companies in Southeast Asia in 1993, 54 were engaged in manufacturing (*Asiaweek*, Chinese language edition, 24.10.93, pp. 50-85, 88-91. Hill, 1988).

Conclusion - Transformation and Continuity

The inaptness of theories, that stressed either the insuperable smallness and dependency of Chinese capitalism, or else its traditional, backward or 'ersatz' nature, was becoming increasingly evident during the 1990s. Arguably two kinds of misunderstanding were involved. The first serously underestimated the characteristics and potential of Chinese businesses. The second failed to take into account the changing nature of the environment presented by the globalising world economy and by technological change.

The supposed inability of Chinese fortunes to grow large was not borne out. Indeed rapid social mobility and many rags to riches stories, within one or two generations, characterise this kind of capitalism and motivate it. The periodic declines and splitting up of such fortunes create space and incentive for new *parvenus*. While few very large companies are produced, there are no inherent obstacles to the accumulation of huge family fortunes which, like Chinese economies as a whole, grow by multiplication of small units rather than expansion of large bureaucratic structures.

One reason for the discrepancy between the wealth of the richest families and the small size of most of their firms is that the retention of family control puts limits on how far public ownership of listed companies can be allowed to dilute the family's direct or indirect share. Another reason was that it made possible the retention of strategic control by family members with long term commitment, who had been educated and prepared for such roles, while professional managers could be allocated administrative functions within the individual companies. The conglomerate structure, and a strategy of diversification, tended to multiply individual units rather than to increase their size. We should be wary, however, of ignoring the economic power of such conglomerates of small firms.

A large proportion of Chinese capitalists remain themselves of small and medium size. Another fallacy is to ignore or deny the extent to which the effectiveness of these can be multiplied by trust based networking amongst firms and their owners. The reach of both large and small companies is extended through often transnational networking between firms. This provides for pooling of information and credit, for reliable supplies and customers, for elaborate sub contracting arrangements and for joint projects. Personal trust, far from being limited (as Fukuyama believes) to a narrow ascribed circle is actively sought and cultivated to an exceptional degree.

Far from trust being restricted to relatives, it is clear that kinship is no more than one amongst a number of bases for the deliberate and active construction of trusting relationships, and active voluntary organisations[17]. Wong argues that in Hong Kong:

> In Chinese economic conduct the crucial distinction is not that of kin and non kin, but personal and impersonal. The Chinese tend to personalise their economic relations. Shared characteristics, including real or fictive kinship, regional origin, educational background etc, often provide convenient bases for cooperation to be initiated. Where such commonalities are lacking, intermediaries are usually brought in as guarantors or brokers to set up the necessary personal linkages... Such particularistic ties are highly elastic... family ties only serve as the nucleus from which a Chinese can spin a web of ever-widening social circles (1988, pp. 136/137).

Our own research indicates that the extent of trust depends on the duration and experience of the relationship, not on the initial closeness of blood. The most trusted long term business associates need not be related at all and may indeed be from other dialect groups or indeed, on occasion, of non Chinese ethnicity (Lever-Tracy et al.,1991, pp. 77-80; Lever-Tracy et al., 1996, Ch. 8).

Chinese business people today, in the Chinese economies of Hong Kong, Taiwan and Singapore and as economically powerful minorities around the world, are well able to use law and contract, and to deal through impersonal institutions. They retain, however, a strong preference for dealing on the basis of personal trust, which is seen not so much as a constraint but as an advantage, and thus they devote a great deal of time and effort to testing and constructing such relationships[18].

Since the adoption of the theoretical blinkers, that had assumed the irrelevance of small Chinese businesses and their distinctive business culture, the world had changed, creating for them a far more promising environment than before. Microelectronics had discounted many of the advantages of mammoth size and mass production. A deregulated global economy had increased uncertainty and competition, replacing the primacy of organisation and predictability with entrepreneurship and flexibility. In the countries of the *core*, commercial capital and buyer chains were gaining prominence over manufacturing capital and producer chains, and were concerned to monopolise marketing and brand names but quite willing to let go of the technology and expertise of production. Multinational corporations had proved unable to develop the *third world* or lead the way to capitalism for the old Soviet bloc, but the opening of China and its surge of decentralised rural industrialisation provided the transnational networks of the diaspora with an opportunity that others had been neither willing nor able to take advantage of.

Does the crisis show all these gains to have been illusory or unsustainable and built on sand? Has it revealed persisting underlying weaknesses in Chinese capitalism that contributed to the crisis or made their businesses vulnerable to its effects and to future recurrences? Has the world environment changed again and is it the case that in the new world of mega-mergers, 'Asia's family-owned conglomerates could face a lonely death if they don't find a partner' (Hiscock, 1999, p32) Does the crisis mean 'The Death of Asian Capitalism' as the title of a 1997 conference on the subject suggested? Or was it just a temporary interruption, in which these same resources might contribute to recovery? These questions will be left to later chapters.

Notes

1   For example *Time, Asiaweek, Far Eastern Economic Review, Forbes, Fortune International, Nomura Asian Perspectives.*
2   The first was held in Singapore and subsequent gatherings were held in Hong Kong, Bangkok and Melbourne.
3   A representative collection of this work can be found in Clegg and Redding (1991).
4   For Yamaguchi (1993, p. 17) this was not so much a sphere that was still struggling to join the industrialised world as one that had in principle gone beyond it, being *information oriented* rather than *technology oriented*, an orientation that combined

personal networking and rapid decision making with modern communications and information technology.

5   In China although inequality within localities and between provinces increased, the largest inequalities, which were between urban and rural areas within the same province, declined after reform (Chai, 1992).

6   This was despite widespread corruption and self enrichment by new elites.

7   For example, by 1996 life expectancy in China was nearly ten years longer than in India and infant mortality was at less than two thirds of the Indian level, although both countries had had roughly comparable figures 20 years earlier (Kurian, 1997).

8   In 1962 Taiwan's per capita income ranked 85th out of 129 countries, between Zaire and the Congo. By 1986 it ranked 38th, between Greece and Malta. South Korea had moved up from 99th, between Sudan and Mauritania, to 44th, between Surinam and Argentina; Hong Kong from 40th, between Spain and Malta, to 28th, between Saudi Arabia and Israel; Singapore from 38th, between Greece and Spain, to 25th, between New Zealand and the Bahamas (Hobday, 1995b. p. 15). Taiwan has overtaken Japan in some areas of electronics production (*FEER*, 11 June 1998, p. 57); In 1978 China ranked 32nd in the world as an international trading economy, in 1992 she ranked 11th (Yan, 1995, p. 10).

9   These figures refer only to the modern, corporate, private sector of the economy: Chinese holdings in listed companies not under state or foreign control, as% of market value of all shares in such firms.

10   This profile is derived from material in *FEER* (7.4.83); *SCMP* (25.6.93 and 8.7.93); *Forbes Magazine* (18.7.94); Yamaguchi (1993, p. 4); *International Business Week* (12.7.93), pp. 41, 44-81; Sato (1993, p. 408), gives a figure of 135 000 employees in 427 affiliated companies for 1990.

11   Greenhalgh (1988, pp. 228, 229) cites research showing that small and medium firms in Taiwan were technically more efficient than large firms in all but one of the fastest growing manufacturing areas in the 1960s, and that they made a disproportionate contribution to export oriented growth.

12   In 1994 ACER became one of the top ten sellers of PCs in the US in 1994. Its US factories assembled components made in Taiwan (*FEER*, 26 January 1995, p. 52).

13   Mackie (1989, p. 97). He argued that this represented a major new development over the previous 20 years. The international links between Chinese businessmen had 'enormously extended the rudimentary trading and credit linkages that were established before World War II'. See also Mackie (1992), pp. 161, 183/4.

14   This is compiled from data on the 200 largest companies in Hong Kong by market capitalisation published weekly in the *Sunday Morning Post* Money Section and data on ownership structure from *Corporate International's Company Handbook* - Hong Kong, various issues (published bi-annually).

15   This figure includes South Korea, which does not feature in our argument, but which provided only US$4.7 billion, around 10% of the total FDI outflows from East and Southeast Asia (excluding Japan).

16   It has been estimated that 97% of Asian investment was financed by Asian savings, with shortfalls in some parts more than covered by surpluses elsewhere (*The Australian*, 2.6.94). Mackie (1992, p. 165) argued that the prevalence of joint ventures around the region in which,

Chinese firms are the most favoured partners or agents of foreign companies - and frequently their competitors - testifies to the great shift that has taken place in the bargaining strength of the two sides. Increasingly it is the foreigners who most need the domestic Chinese in order to get access to Southeast Asian markets, whereas the big Chinese corporations are increasingly able to pick and choose worldwide in obtaining the capital and technology they need from foreign companies. The Globalisation of capital over the last thirty years has worked very much to their advantage.

[17] On the prevalence and importance of Chinese organisations see for example Wickberg (1988); See (1988); Omohundro (1981, Ch 4); EAAU (1995, pp. 16/17, 288-309).

[18] It is precisely because trust is central to the successful operation of Chinese economies that mistrust and the recognition of the need to test and build trust is also widespread. On the importance to Chinese of developing and achieving trust see, for example, Hwang (1987); Inglis (1974); Redding (1990); Lim and Gosling (1983) on Chinese in Southeast Asia; Chan and Chiang (1994 pp. 59-62) on Singapore; Omohundro (1981 especially pp. 62-69, 113), on the Chinese in the Philippines; Numazaki (1991, p. 89) on Taiwan; Wong (1988, 1990) on Hong Kong; Lever Tracy et al (1991) on Chinese in Australia.

References

Asiaweek.

Beijing Review.

Beng, C. S. (1988), *Small Firms In Singapore*, Oxford University Press, Singapore.

Borrus, M. (1994), 'Left for Dead: Asian Production Networks and the Revival of US Electronics', in Eileen M. Doherty (ed.), *Japanese Investment in Asia: International Production Strategies in a Rapidly Changing World*, The Asia Foundation, BRIE, San Francisco.

Chai, J. (1992), 'Consumption and Living Standards in China' *The China Quarterly*, 131, September.

Chan, K.B. and Chiang, C.S.N. (1994), *Stepping Out: The Making of Chinese Entrepreneurs*, Prentice Hall, Singapore.

China's Customs Statistics, Monthly.

Chinben, See (1988), 'Chinese Organisations and Ethnic Identity in the Philippines', in Cushman J. and Wang, G. (eds).

Chiu, P.Y.W. (1988), *The Economy Of Hong Kong*, Enterprise Publishing, Hong Kong.

Christerson, B. and Lever-Tracy, C. (1997), 'The Third China? Emerging Industrial Districts in Rural China', *International Journal of Urban and Regional Research*, 21/4, December.

Clegg, S. and Redding, S.G. (eds) (1990), *Capitalism in Contrasting Cultures*, 1990, Walter de Gruyter, Berlin.

Corporate International's Company Handbook.

CSY (China Statistical Yearbook).

Cushman, J. and Wang, G. (eds) (1988), *Changing Identities of the Southeast Asian Chinese Since World War ll*, Hong Kong University Press, Hong Kong.

DOTS (Direction of Trade Statistics), Yearbook. IMF (International Monetary Fund).

EAAU (East Asia Analytical Unit) (1995), *Overseas Chinese Business Networks in Asia*, Department of Foreign Affairs and Trade, Parkes, Australia.

Evans, P. (1987), 'Class, State and Dependence in East Asia: Lessons for Latin Americanists', in Deyo, F.C. (ed.), *The Political Economy of the New Asian Industrialism*, Cornell University Press, Ithaca.

FEER (Far. Eastern Economic Review).

Forbes Magazine.

Fortune International.

Fukuyama, F. (1985), *Trust: The Social Virtues and the Creation of Prosperity*, Hamish Hamilton, London.

Greenhalgh, S. (1988), 'Families and Networks in Taiwan's Economic Development' in Winckler E. and Greenhalgh, S. (eds), *Contending Approaches to the Political Economy of Taiwan*, An East Gate Book, Armonk.

Hamilton, G. (ed.) (1991), *Business Networks and Economic Development in East and Southeast Asia*, Centre for Asian Studies, University of Hong Kong.

Hill, H. (1988), *Indonesia's Industrial Transformation*, Allen and Unwin, Singapore.

Hiscock, G. (1999), 'Merge or perish', *The Australian*, 26th April.

Hobday, M. (1995), *Innovation in East Asia: The Challenge to Japan*, Edward Elgar, Cheltenham.

Hong Kong Digest of Statistics (1991).

Hsiao, H.H. and M. (1989), 'The Middle Classes in Taiwan: Origin, Formation and Significance' in Hsiao, H.H., Cheng, W.Y. and Chan, H.S. (eds), *Taiwan, a Newly Industrialised State*, Proceedings of 1987 conference held in Department of Sociology, National University Taipeh.

Hsing, Y.T. (1998), *Making Capitalism in China: The Taiwan Connection*, Oxford University Press, New York.

Hwang, K.K. (1987), 'Face and Favor: The Chinese Power Game', *AJS* Vol 92/4 January.

IMF (International Monetary Fund), *Direction of Trade Statistics.*

Inglis, C. (1975/6), 'Particularism in the Economic Organisation of the Chinese in PNG', *Anthropological Forum*, Vol IV/1.

International Business Week.

Kojima, S.(1994), 'Alternative Export-Oriented Development Strategies in Greater China', *Jetro China Newsletter*, No. 113, November-December.

Kotkin, J. (1993), *Tribes: How Race Religion and Identity Determine Success in the New Global Economy*, Random House, New York.

Kurian, G.T. (1997), *The Illustrated Book of World Rankings*, M.E. Sharpe, Armonk.

Lever-Tracy, C., Ip, D. and Tracy, N. (1996), *The Chinese Diaspora and Mainland China: An Emerging Economic Synergy*, Macmillan, Houndmills.

Lever-Tracy, C., Ip. D., Kitay, J., Phillips I. and Tracy, N. (1991), *Asian Entrepreneurs in Australia: Ethnic Small Business in the Chinese and Indian Communities of Brisbane and Sydney*, OMA/AGPS, Canberra.

Lever-Tracy, C. and Tracy, N. (1993), 'The Dragon and the Rising Sun: Market Integration and Economic Rivalry in East and Southeast Asia', *Policy Organisation and Society,* Summer.

Lim, L.Y.C. (1983a), 'Chinese Economic Activity in Southeast Asia: An Introductory Review', in Lim and Gosling.

Lim, L.Y.C. (1983b), 'Chinese Business, Multinationals and the State: Manufacturing for Export in Malaysia and Singapore', in Lim and Gosling.

Lim, L.Y.C. and Gosling, L.A.P. (eds) (1983), *The Chinese in Southeast Asia, Volume 1 Ethnicity and Economic Activity*, Maruzen Asia, Singapore.

Limlingan, V.S. (1986), *The Overseas Chinese in ASEAN: Business Strategies and Management Practices*, Vita Development Corporation, Manila.

Lin, P.A. (1991), 'The Sources of Capital Investment in Taiwan's Industrialisation' in Gary Hamilton (ed.), *Business Networks and Economic Development in East and Southeast Asia*, Centre for Asian Studies, University of Hong Kong.

Mackie, J. A.C. (1989), 'Chinese Businessmen and the Rise of Southeast Asian Capitalism', *Solidarity*, 123, July/September.

Mackie, J.A.C. (1992), 'Overseas Chinese Entreprenurship', Asian-Pacific Economic Literature, Vol 6/1, May.

McVey, R. (ed) (1992), *Southeast Asian Capitalists*, SEAP, Cornell University, Ithaca, NY.

Nikkei Weekly.

Numazaki, I. (1991), 'The Role of Personal Networks in the Making of Taiwan's Guanxiqiye' (Related Enterprises), in Hamilton (ed).

Omohundro, J.T. (1981), *Chinese Merchant Families in Iloilo. Commerce and Kin in a Central Philippine City*, Ateneo de Manila University Press, Quezon City.

Pan, L. (1990), *Sons of the Yellow Emperor*, Secker and Warburg, London, 1990.

Practices, Vita Development Corporation, Manila.

Rabushka, A. (1987), *The New China: Comparative Economic Development in Mainland China, Taiwan and Hong Kong*, Westview Press, San Francisco.

Redding, S.G. (1990), *The Spirit of Chinese Capitalism*, Walter de Gruyter, Berlin.

Redding, S.G. (1991), 'Weak Organisations and Strong Linkages: Managerial Ideology and Chinese Family Business Networks' in Hamilton (ed).

Sato, Y. (1993), 'The Salim Group in Indonesia: The Development and Behaviour of the Largest Conglomerate in Southeast Asia', *The Developing Economies* XXXI/4 (December).

SCMP (South China Morning Post), various dates.

Seagrave, S. (1995), *Lords of the Rim: The Invisible Empire of the Overseas Chinese*, Bantam Press, London.

Sit, V.F.S and Wong, S.L. (1989), *Small and Medium Industries in an Export-Oriented Economy: The Case of Hong Kong*, University of Hong Kong, Hong Kong.

STW (Straits Times Weekly), various dates.

Sunday Morning Post, various dates.

SWB (Summary of World Broadcasts), various dates.

Tam, S. (1990), 'Centrifugal Versus Centripetal Growth Processes: Contrasting Ideal Types for Conceptualising the Development Patterns of Chinese and Japanese Firms', in Clegg and Redding (eds).

The Australian, various dates.

The Economist, various dates.

UN (United Nations), *World Investment Report (annual)* various dates.

Weidenbaum, M. and Hughes, S. (1996), *The Bamboo Network: How Expatriate Chinese Entrepreneurs Are Creating a New Superpower in Asia*, The Free Press, New York.

Wickberg, E. (1988), 'Chinese Organisations and Ethnicity in Southeast Asia and North America since 1945: A Comparative Analysis', in Cushman and Wang.

WIR (World Investment Report, annual), United Nations.

Wong, G. (1991), 'Business Groups in a Dynamic Environment: HK 1976-1986', in Hamilton, 1991.

Wong, S.L. (1988), 'The Applicability of Asian Family Values to Other Sociocultural Settings', in Berger, P. L. and Hsiao, H.H.M. (eds), *In Search of an East Asian Development Model.*

Wong, S.L. (1990), 'Chinese Entrepreneurs and Business Trust', University of Hong Kong, *Supplement to the Gazette*, Vol. XXXVII/1, May.

Wu, F. (1997), 'Hong Kong and Singapore: A Tale of Two Asian Business Hubs', *Journal of Asian Business*, 13/2.

Wu, Y.L. and Wu, C.H. (1980), *Economic Development in Southeast Asia the Chinese Dimension*, Hoover Institution Press Stanford, 1980.

Yamaguchi, M. (1993), 'The Emerging Chinese Business Sphere', *Nomura Asian Perspectives*, Vol. 11/2, July.

Yan, S. (1995), 'Export Oriented Rural Enterprises', *Jetro China Newsletter*, 118 September-October.

3   The East Asian Financial Crash: Causes and Consequences

NOEL TRACY

When I first envisaged this chapter, it was to explain the financial and economic crisis, that began in mid-1997 and rapidly enveloped the whole of East Asia; to explain it in the context of financial crises in general; and to consider the impact of financial crises on the structure of economies. Part of that agenda has been somewhat changed by the rapid turnaround in both the economies of the region and in their prospects as we enter the new millenium. In the last days of 1999, stock markets in both Hong Kong and Singapore closed at record levels, both having registered gains in excess of 60% in the year, while at the same time forecasts for economic growth for much of the region for 2000 were being constantly revised upwards.

While there can be little doubt that substantial problems remain - Japan remains in the doldrums, South Korea's Chaebol and banks remain locked in a bad debts/bad loans embrace, Thailand and Malaysia are far from out of the woods, Indonesia's recovery has barely started and China's problems in its state-owned sector and banking system remain seemingly intractable - confidence has returned to much of the region's economy. My task is, therefore, not only to explain the crisis of 1997-98 but also to explain the very rapid recovery, particularly in financial markets[1], from what was painted by many commentators as the most serious economic crisis since 1945 and the end of the East Asia Miracle, that began and quickly gained pace in the second half of 1999.

When the crisis first broke in mid-1997, initial analyses concentrated on failings within the economies, polities and societies of East Asia. Critics pointed to weak banking systems, poor financial regulation and above all to 'crony capitalism', corrupt linkages between business and government that had led to 'moral hazard' for both lenders and borrowers as it became increasingly difficult to judge what was private and what was sovereign risk[2]. This kind of thinking was clearly behind the IMF 'rescue packages' that were imposed on Thailand, Indonesia and South Korea: governments needed to be forced to impose order and restraint on their financial markets and to end crony privileges and linkages particularly in government contracts and in access to credit. The weaknesses of this kind of analysis and the remedies it proposed, which did not detract from their attractiveness in certain quarters, was that none of these failings were new, indeed all had been in place during the boom periods, and if anything had been somewhat improved in the period immediately prior to the crash.

This type of analysis was undermined from a number of quarters. Firstly by commentators who suggested that, while this might indeed be a reasonable explanation for the crash in a number of countries, particularly Indonesia, Malaysia, Thailand and South Korea, it didn't really provide an explanation as to why the crisis was region-wide. Second, it provided no explanation of why it had affected the whole region so quickly and abruptly[3]. Third, whatever the failings of the banking systems in other parts of the region, these clearly did not apply in Hong Kong and Singapore, their banking systems being noted for their tight regulation, prudence and high capital adequacy. Nonetheless both had been particularly badly affected by the crisis very quickly and subject to substantial falls on their stock markets and marked pressure on their currencies. Localised explanation were plainly inadequate to explain a crash of such depth and breath as had engulfed the whole East Asian region by the end of 1997.

What shifted the balance towards other explanations was clear evidence that the IMF packages were not only not working but were counter-productive. This was demonstrated in the social turmoil in Indonesia and the collapse of the Suharto regime, and the spread of the crisis first to Russia in August and September 1998, leading to defaults in both foreign and domestic debt, and to Brazil by January 1999. In between, the collapse of the US Hedge Fund, Long Term Credit Management, had required a bailout by its international creditors totaling US$3.3 billion. By the second half of 1998, few commentators considered the crisis a purely East Asian phenomena or even a crisis which could be quarantined to East Asia. From this point on, US and IMF policies were concentrated on maintaining domestic and international liquidity, reducing and keeping interest rates low and encouraging a return to fiscal and monetary stimuli in the affected economies. While this change of policy was not of itself sufficient to ensure recovery, it made recovery possible by putting downward pressure on interest rates and encouraging monetary growth and loosening credit. Following this sea change in official attitudes, early 1999 saw the first signs of recovery and the latter part of the year saw a return to growth in both domestic economies and regional trade combined with a rapid return of confidence in financial markets. By the end of the year, Hong Kong and Singapore's stock markets were at record levels, new stock issues were again attracting attention and over subscription and there were substantial recoveries in stock markets as well as the overall economy all around the region.

How then are we to begin explaining the crisis and the rapid recovery? Clearly the initial explanation of inadequacies in financial structures around the region will not do. The recovery has come long before these have been adequately remedied, much still needs to be done in Indonesia, Malaysia, Thailand, South Korea, China and Japan. Nor can these long-standing problems give us an explanation for the speed and rapidity with which the crisis engulfed the whole region. There can be little doubt that these problems and inadequacies contributed to the crisis, made it worse and harder to resolve but they cannot be a sufficient cause of the crisis. Any adequate explanation must account for the sudden collapse right across the region. In order to provide this we need both a general theory of financial crashes and a specific analysis of how this applied in this crisis.

In seeking this, I was drawn to two long neglected and unfashionable works by Charles Kindleberger, *Manias, Panics and Crashes: A History of Financial Crashes* (1978) and *The World in Depression 1929-39* (1973). The first point Kindleberger makes is that financial crises are associated with peaks of business cycles and with the increase in speculative activity that produces:

> Speculative excess, referred to concisely as mania, and revulsion from such excess in the form of a crisis, crash or panic can be shown to be, if not inevitable, at least common (1973, p.4).

In his analysis of financial crashes since the South Sea Bubble in 1720, Kindleberger drew heavily on the work of a monetary theorist, Hyman Minsky, noted for his pessimistic views on the fragility of the monetary system and its propensity to disaster (1978, p.15). Minsky's monetary model had particularly emphasised the instability of the credit system. Booms are fed by increases in bank credit but where this becomes dangerous is when speculation for price increases is added and threatens to replace investment for production and sale as the major reason for borrowing. The 'euphoria' this produces can lead to 'overtrading', an increase in speculative trading produced by overestimates of prospective returns and by excessive gearing, that is buying on margin in the expectation that it will be possible to make a profit out of onselling the asset while passing on the future obligations to make the balance of the payments. While professional investors frequently do this, it becomes a 'speculative bubble' when sections of the general population are drawn into such activities on a large scale (ibid., p.17).

The 'bubble' bursts when there is a growing awareness that prices have peaked and it is time to get out. Anything can trigger this from a business collapse to a spate of rumours. This turns into a panic when suddenly there are vastly more sellers than buyers and prices start to fall sharply. What had been investment 'euphoria' overnight turns into 'revulsion' with no buyers and everyone seeking to get out of assets into cash. Panic then ensues with prices collapsing, investors selling at a loss and those left holding assets not only incurring huge losses but often being unable to service even their debts, leading to defaults (ibid, pp. 19-20). Kindleberger suggested the model was particularly applicable to international financial and currency markets. In these markets, unlike the domestic economy, the degree of control governments can exercise to restrain excess are severely limited.

How does this model fit the Asian crash? Very closely has to be the answer. Clearly investor 'euphoria' had taken over in the years prior to the crash. Speculation on price increases/capital appreciation by domestic investors had increased markedly. This was particularly noticeable in the real estate sector. This was brought home to me in Hong Kong in the latter months of 1996. In the expectation of an economic bonanza following reunion with China the following year, domestic investors were chasing property under the illusion that prices had only one way to go and there would soon be a substantial new market for luxury office and residential accommodation from mainland companies and staff. The result was that any development attracted buyers prepared to pay almost any price to get a piece of the action. Development on the Kowloon side, promoted as luxury accommodation that would have attracted little interest ten years before, involving sometimes quite modest apartments for sale at US$1million and more, attracted queues as soon as they came onto the market. The speculative mania involved became apparent when one developer offered potential buyers, who had formed a queue a week before the properties became available for sale, numbered tickets so that they could resume their places without having to camp out for a week. These tickets were selling for up to HK$250,000 in the following week[4]. The result of this and similar activities all over the colony was an increasing number of people investing in the property market on the assumption of either exceptionally high returns or capital gains from asset inflation as prices continued to be driven up by excessive demand. On the other side of the equation, people of often quite modest means were borrowing beyond what they could service from their own salaries and business incomes in the expectation of being able to pay the interest on the loans from rental income before they realised their capital gain.

While the example described above might be seen as an exceptional case, the reality was that there were speculative property booms all around the region, in Singapore, Kuala Lumpur, Bangkok, Jakarta and even in Chinese cities like Shanghai, Beijing and Guangzhou, largely fuelled by developers building new office complexes and private buyers investing in rental property rather than buying for their own use. The result was the highest office rents in the world and residential rents that could only be paid by those with substantial employer-funded housing allowances[5]. Governments added to the speculative frenzy as they supported and sponsored prestige projects like Kuala Lumpur's Twin Towers and high-tech corridor, Cyberjaya.

Much of this speculative activity was funded by an increase in bank credit. This was made possible by an enormous increase in private bank lending by international banks into the region. Financial deregulation had enabled many international banks to open branches in the region. These were now competing with established local banks for customers[6]. Other international banks had increased their interbank lending to local banks. It is interesting to note how this had impacted upon the five most seriously affected countries in the 1997-98 crisis (Thailand, Indonesia, Malaysia, Philippines, South Korea). In 1994, international commercial banks loaned some US$23.4 billion to banks, corporations and private individuals in these economies. Non-bank financial organisations had loaned a further US$2.4 billion. Two years later, this had increased to US$55.7 billion and US$22.7 billion respectively, an increase of over 200 % (Institute for International Finance, 1998).

Prior to the 1990s, the region, as a result of the high savings of its citizens, had largely been able to finance its own investment needs. The boom of the 1990s, however, had led to an increase in borrowings as capital requirements had exceeded domestic savings in many countries. This change in the structure of borrowing, allied with the fact that much of this new borrowing from international sources was short-term, had dramatically increased the vulnerability of these economies to external shock. This shock came in 1997, when a withdrawal of funds from Thailand in the face of a declining current account position, led to the sudden withdrawal of international funds from the region as international banks and investment funds started to reassess their risk. Commercial banks which had loaned more than US$55 billion to the five most effected economies in 1996, withdrew almost US$27 billion in 1997 (ibid), a turnabout of more than US$80 billion in a single year equivalent for these five economies of losing 10 % of their GDP.

Economies which had been flush with easy credit suddenly found there was no credit to be had, as banks called in loans and refused to make further advances. The vulnerability of these economies not only resulted from the increase in international borrowing but from the fact that most of this was short term. Not only were bank clients borrowing for speculative purposes but they were borrowing short-term funds thereby multiplying their chances of being caught out by any change in the business climate never mind the full scale panic that ensued in the second half of 1997.

The Asian financial crash of 1997, therefore, follows a definite pattern: a boom engendered by easy credit, which develops into a speculative mania as more and more people seek to make their profits from asset appreciation rather than production and sale, which suddenly comes to an end when panic sets in and everyone becomes asset averse and seeks to move into cash. In this case it was less investors than lenders who were first to panic and rush to get out.

The rush out of bank lending was parallelled by international fund managers selling off their share portfolios as they too tried to get out before prices collapsed. While portfolio investment from international funds had increased substantially in the 1990s, it was on nothing like the same scale as lending by international banks. It nonetheless still took its toll: portfolio funds had bought US$12.4 billion in local equities in 1996, it 1997 this was reversed into net sales of US$4.3 billion. The impact of this selling pressure on stock markets was dramatic: prices fell through the floor.

The withdrawal of funds from the banking sector, real estate and stock markets was paralleled in the currency markets. Banks calling in their loans and fund managers selling their shares were also selling local currencies to repatriate their money. Local corporations and private individuals, who had borrowed in foreign currency, were also forced to sell local currencies in order to repay their foreign debt as quickly as possible, as local currencies plunged. The crisis was, therefore, compounded by asset sales, repatriation of foreign funds and panic selling of local currencies. As currencies fell, Hedge Funds moved in: short selling, in the expectation of making profits on further falls in the value of the respective currencies, a prospect that their activities turned into a self-fulfilling prophecy. This analysis, therefore, clearly locates the origins and the ongoing causes of the Asian financial crisis in the problems and weaknesses of international financial markets rather than in the domestic economies of the region. This is not to deny the weaknesses of much of East Asia's financial architecture nor its part in the crisis but to locate the principal causes of the financial crash first in the rapid build up of international financial funds in the region leading to a speculative orgy and then in the sudden collective panic withdrawal of funds once the crisis broke.

What are the problems and weaknesses in international financial markets that led to this debacle? First, we need to recognise that the amount of money currently parked in savings accounts, investment accounts, pension funds, mutual funds now far exceeds the profitable investment opportunities available. This leads to an increasing search for short-term profits by the financial managers charged with controlling these funds, who must justify their position not only in terms of the profitability and return of the funds they control but in comparison with other fund managers. The result is an enormous amount of 'hot money' endlessly searching for quick returns. The sheer volume of this 'hot money' is sufficient to sink virtually any convertible currency or stock market it subjects to a sell off.

The enormous growth of deposits at their disposal has led to increasing competition among banks to lend[7]. This increasing competition to lend has been accompanied by an equally rapid internationalisation of banking operations. As banks internationalise their operations, opening new branches in a variety of locations in which they have had no previous presence or experience, they must seek to attract customers away from existing well-established banks in those locations. They can only do so by offering easier terms and by financing projects and borrowers of which the established banks are wary. As more and more banks open and compete with each other in any location, the local banking expertise is spread thinner and thinner. The result was more money available to lend, increasing competition to lend and a thinner spread of real expertise, a disastrous cocktail which invariably led to greater risk taking both with projects and clients to be funded. Local banks, which had previously pursued fairly conservative lending policies, were forced to match the lending of newly-arrived international banks or lose their customers. To compound the problem, local banks were also borrowing from international banks in order to increase their liquidity and thus capacity to lend.

The combination of new international banks, plus an increase in interbank lending to local banks from international sources, with increasing involvement in local equity markets by international fund managers, dramatically increased the liquidity of the region's economies. They did so, however, substantially by a massive build up in the volume of short-term money, 'hot money', rather than by long-term investment in the region.

In this situation the lack of adequate financial regulation became critical. Many of the region's economies had only deregulated their financial sectors since the 1980s. This applied in particular in Indonesia, Malaysia, Thailand and South Korea. Hong Kong and Singapore, by contrast, had maintained substantial controls and also had adequate experience of policing both the banking system and foreign banks. Other economies, most notably China and Taiwan still had substantial restrictions on capital account transactions and were thus less vulnerable to movements of hot money and excessive competition among banks.

It might be asked at this point why governments had been so lax and complacent about this massive build up of short-term money. The problem was that in the period of high growth from the late 1980s onwards, Japan excepted, these funds were needed to fuel the growth of the economy. These funds increased bank liquidity, enabled banks to finance corporate expansion, financed share floats enabling entrepreneurs to raise funds for expansion and generally buoyed up stock markets. They also enabled grandiose government projects to proceed without draining the markets of liquidity and squeezing out private sector activity. Domestic savings were by this time insufficient to cover this rapid increase in the demand for money.

On the eve of the crisis, we, therefore, had a massive build up of over-borrowing by government and private corporations and individuals matched by an orgy of overlending by both local and international banks. With no restrictions on the movement of capital, once panic set in and everyone sought to liquidate their exposure, the flight of capital soon became a flood and liquidity quickly disappeared from markets. In addition, domestic institutions, banks, corporations and individuals sought to buy foreign currency to cover their exposure to offshore loans denominated in most cases in US dollars. Soon there was nothing holding Asian currencies up in international markets and the most exposed currencies, the Thai Baht, Indonesian Rupiah, Malaysian Ringit, and South Korean Won tumbled by over 50% in rapid succession. The collapse of the Japanese Yen to around 147 to the US dollar (from 90 in 1995) added to the rout.

Markets that had been flush with liquidity only months before were now starved of liquidity. Corporations found that credit could not be had at any price. Loans that had been committed to projects were now called in, in many cases without notice. Corporations and individuals that had borrowed in foreign currency now found themselves technically insolvent as their assets denominated in local currency and depreciated accordingly in value were now often insufficient to cover their debt. Corporations that had contained substantial shareholder equity now found themselves technically in the hands of their creditors. Stock markets now found themselves subject to 'revulsion', turnover had collapsed[8] and buyers were few.

The problem for local banks was that, with many of their clients now technically insolvent, they were sitting on a mountain of bad debt and with their liabilities often exceeding their performing assets were themselves in many cases also technically insolvent. This was widespread in Thailand, Indonesia, Malaysia and South Korea, with most local banks in need of massive injections of new capital. With the easy money from international banks no longer available, they either had to introduce new capital from their own sources, owners or shareholders, raise new capital from local or international markets or seek salvation from their governments through central bank intervention. Raising new capital even on extremely onerous terms in conditions of 'revulsion' was no easy task. Owners and shareholders were often stretched by personal borrowings and by asset depreciation to the point were raising new capital was difficult, while at the same time, neither local nor international banks were lending, including even those among the former who still had the means to do so. Central Bank intervention was in many cases the only available means of keeping them afloat at least in the short term. IMF terms for providing the funds to enable this to occur required substantial reform and regulation of the banking system. In the worst hit cases, the governments of Thailand, Indonesia and South Korea found they had no choice but to seek IMF assistance regardless of the conditions. Malaysia managed to struggle through without such intervention, but at enormous cost, and was forced to re-introduce capital controls in September 1998.

One of the advantages of the explanation of the Asian Financial Crisis in terms of the problems of the international financial system, in particular the enormous growth of 'hot money' and the instability both potential and real this introduces, and in terms of panic as the bubble burst in 1997, is that it also provides us with an explanation of the sudden comeback of the region's financial markets in the second half of 1999.

If the weaknesses of local financial institutions and markets were the principal cause of the debacle of 1997, explaining the recovery would be quite difficult. Reform of the financial sector in Indonesia, Thailand, Malaysia and South Korea has only just begun and is a long way from completion. In Indonesia and Thailand, many banks are effectively under central bank and state control as the previous owners have neither the new capital required to recapitalize them nor the confidence of the authorities in their capacity to manage them in future. New buyers and investors are being actively sought but this is only proceeding slowly and most disposals have been of the smaller institutions. The real test of this policy will come when larger institutions are readied for sale. Malaysia had sought to consolidate its banks into six large groups but this policy failed in the face of opposition from existing bank owners and the very poor politically motivated choice by the government of some of the core banking entities around which the new structure was to be built[9]. In addition, the financial sectors in Japan, China and Taiwan are far from secure, even though they avoided the indignities of IMF intervention during the crisis. Banks in both Japan and China are overburdened with bad and, in many cases, irrecoverable debts. The structure of the banking system and its management is suspect in both cases. Taiwan's heavily protected banks are far from internationally competitive and have a very poor record of lending to emerging companies, the growth sector of the economy over the past four decades.

Clearly recovery in financial markets has not come because the problems in the region's financial institutions have been overcome and remedied. This process, if it is indeed ever completed, still has a long way to go. What has clearly buoyed local markets is the return of international funds in 1999 both in the form of bank lending and portfolio investment (Institute for International Finance, 1999). International banks have begun lending to the region again and fund managers have seen value in local stock markets. The panic of 1997 had seen a rush of both banks and fund managers back to the best performing economy of the second half of the 1990s, the United States. However, despite, the strong performance in 1999 on both exchanges, Wall Street and Nasdaq, Fund Managers, in particular, were beginning to see themselves as dangerously overexposed to a single market. Continuing jitters about how long the US bull market could continue also inclined managers to seek to spread their risk. Fears about interest rates, spurred by statements from Alan Greenspan, Chairman of the Federal Reserve, about the need to hose down overheating, added to the concern. The continuing blowout in the US trade deficit, which reached US$300 billion in 1999, also raised concerns about the level of the US dollar. In this situation, with the worst in Asia clearly over, the better regional stocks once again became attractive, particularly at the low price levels in the first half of the year.

By the end of the year, Asian stock markets, particularly the two strongest markets, Hong Kong and Singapore, were again attracting attention from fund managers with money to invest and risks to hedge. Strong recovery in financial markets, well beyond what was justified by the economic fundamentals, it might be added, was soon the order of the day. By year end, new records were being set. This was a remarkable turnabout in such a short time, less than two and a half years, from the prognoses of the end of the Asian Miracle that were so widespread in 1997.

What impact do financial crises have on the overall structure of economies. The answer is substantial. Financial crises, particularly those as severe as those of 1997 and 1929, while they may occur because of the excesses at the height of the business cycle, are more than just the start of troughs from which growth and business as usual will recommence at some later date. They indicate a generalised crisis after which nothing will be ever quite the same again. International funds may be returning to only partially reformed East Asian financial markets, but the impact on the structure of the economy of the region has been in many ways much more substantial than in financial markets.

The Chinese word for crisis is made up of two characters, one representing danger, the other opportunity. This is a very good way to look at it. A crisis of this magnitude means disaster for some sectors of the economy, some industries, some regions, but new opportunities for others. At the end of the day, the structure of the economy is transformed.

To illustrate this, we can examine the changes in the US economy brought about by the Great Depression. The US was the most badly effected by this crisis and was still in fundamental recession in 1939, fully ten years after the crisis broke and long after many of its competitors had recovered. Japan, by contrast, had substantially recovered by 1933. Despite the length and depth of the crisis, substantial transformation of the economy had taken place. In 1929, heavy industry, much of it located in the Northeast and East Coast, and Chicago bankers were the dominant sectors of the economy. By 1939, the economy had shifted westwards and southwards, new industries like petrochemicals and those supplying consumer demand like consumer durables, particularly household appliances, had grown up. The whole notion of the American dream had changed from either the family farm for the poor or sophisticated New York for the middle classes towards the California lifestyle and professional employment in the new West Coast industries. The movement of people and industry to the West Coast also meant that the US was now a fully fledged Pacific power, with the industrial might in place to fight a war after Pearl Harbour in 1941. By the end of the war, California was the most populous state in the Union and Los Angeles had replaced New York as the quintessential American City.

What has happened in East Asia as a result of the crisis of 1997-1999? The most obvious casualty is Japan. The Japanese economic downturn that began in 1991 still shows no signs of ending and has intensified as a result of the crisis. The banking crisis that ensued laid bare much of the structural weaknesses that beset the Japanese economy. Japan, which 10 years ago was seen as typifying Asian economic might and seemingly unbeatable, has been shown to have feet of clay and no capacity either to solve its own problems or provide regional leadership in the crisis. Clearly, Japan is the big loser from the crisis. The Japanese crisis also indicates where much of the structural crisis and the declining sectors of the regional economy can be located.

It is by comparing the Japanese economy with the American economy that we can most clearly see where the problems lie. If we look at the transformation that has taken place in the American economy since the Oil Crisis of 1973, we can see immediately that the economy is dominated by new players. It is Microsoft, Intel, Sun Microsystems and Cisco Systems which typify the new booming economy in the 1990s with new software and ecommerce stocks regularly listing on the Nasdaq. These new giants didn't even exist in 1973, many are less than 20 years old. While the US corporate structure has been radically transformed by these dynamic new entrants, the same is not true in Japan. where the old corporate giants continue to dominate the economy. While the US economy and society have permitted and even encouraged these new entrants, those of Japan have not. The result is the Japanese economy is still dominated by yesterday's industries, motor vehicles, electronics particularly consumer electronics, household appliances and by the giant construction companies while at the same time their productive capacity is well beyond market demand - markets for these products are already saturated.

By contrast, the US economy is into software, ecommerce and the internet revolution, where demand is constantly rising. The problem for Japan is that neither its corporate structure nor its society are suited to the new economy. The new economy has been produced by risk takers, individuals prepared to stake their future on pursuing what at first was no more than a bright idea and who dreamed of setting up on their own account in order to turn these ideas into world-beating products. Silicon Valley is full of these small companies, Japan has no equivalent. Individualism and non-conformism are frowned upon, making these kinds of developments and changes in Japan problematic. Even 'software entrepreneurs' in Japan like Masayoshi Son (Softbank) and Yasumitsu Shigeta (Hikari) are more shrewd investors in other people's products than genuine IT innovators. Even Japanese banks, which internationalised on a grand scale in the last two decades are now in full retreat. While they provided a large part of international finance in this period, they never succeeded in establishing themselves as market leaders.

Japan is, therefore, stuck with a late-Fordist economy, dominated by industries which are all suffering from overcapacity. This overcapacity is not just local but mirrors similar problems both regionally and globally. Nowhere is this clearer than in its once mighty car industry. The industry globally has massive overcapacity, almost all producers are working at less than full capacity, profitability is poor and falling, and mergers are seen as the best way of reducing the cut-throat competition.

The problem is that substantial parts of the regional economy has been modeling itself on and seeking to follow the Japanese path of industrial development. South Korea has become almost a Japanese clone and as a result become one of Japan's major competitors in international markets. South Korea's corporate debt crisis largely sprang from the rush to expand by the Chaebol in the 1990s and the massive borrowing this entailed in order to take advantage of Japan's declining competitiveness, or so they thought, as the yen appreciated to dizzy heights in the early 1990s. With the collapse of the yen in 1997, this policy had clearly come unstuck. Elsewhere in Asia, many governments saw Japan and South Korea as the preferred model for economic development and corporate structure. The reform of the state sector in China, for example, was to be modeled on South Korea as the government sought to re-organise the state sector into large groups. Malaysia's Look East Policy was similarly modeled on Japan and South Korea. The result is that in many parts of the region, interventionist governments were seeking to build industrial economies following the Japanese model. National cars, steel, chemicals, construction and consumer durables, all received substantial investment and protection and in the process the substantial overcapacity in these basic industries was accentuated, which now threatens all these sectors with downsizing and financial losses.

By contrast other parts of the regional economy, of which Taiwan is the most important example, were moving in another direction. Turning their backs on the Japanese model, they were moving into hi-tech and IT. In an economy in which small entrepreneurial companies had been the movers and shakers since the 1960s, moving into these new sectors was less of a problem than it was in Japan and South Korea. Risk takers abounded. Small business was used to raising capital other than through banks by mobilising the savings of family, friends and neighbours. Many companies had experience of working as OEM and ODM for American computer companies. Taiwan was also the latter's most important source of computer parts and components. Many technicians had been educated in US universities and had served their time in Silicon Valley before coming home to start on their own. The transformation of the Taiwanese economy towards hi-tech and IT was certainly underway before the financial crisis broke but it speeded up the process of restructuring. A number of Taiwanese IT companies even took advantage of the crisis to establish Research and Development facilities in China and recruit previously underemployed computer engineers from the mainland to staff them[10]. A number of Taiwanese software companies have also listed on the Nasdaq in 1999. While Taiwanese industry was always well behind Japan in industrial technology, there is some evidence that its corporations and entrepreneurs have now leaped ahead of Japan and South Korea in the information technology stakes, in particular in the ability to develop software.

Likewise, considerable change has occurred in Hong Kong. Traditionally dependent on the property market and international trade, its entrepreneurs are now moving rapidly into mobile communications and ecommerce. Even staid property developers like Sun Hung Kai and Henderson Land have launched IT ventures, formed joint-ventures with US software giants and sought to introduce IT and hi-tech features into their property developments. Singapore is also seeking to move away from its position as a high-valued added manufacturing base for Multinational Corporations and encourage its IT, hi-tech and software entrepreneurs. At the same time it is upgrading its infrastructure to accommodate the new industries.

While the above analysis suggests it is the Chinese societies that have been most able to take advantage of the crisis, it is important to realise that it is far from a unilinear process. Many Chinese-owned businesses and entrepreneurs have suffered disastrously from the crisis. While many in Taiwan, Hong Kong and Singapore have been able to use the crisis to restructure their business activities, many others in Southeast Asia have not been able to do so and have suffered very badly from the crisis. Nowhere is this clearer than in Indonesia, where many of the largest conglomerates, which had dominated corporate Indonesia before the crisis, are now insolvent and at the mercy of government restructuring agencies which have provided the temporary capital needed to keep them afloat. While the Indonesian-Chinese conglomerates are the biggest losers from the crisis among regional Chinese business[11], their plight is mirrored elsewhere in Malaysia, Thailand, Philippines and Hong Kong, where overborrowed corporations and entrepreneurs have found themselves staring bankruptcy in the face. This was particularly true in the property sector, traditionally a highly-geared sector, wherein not only were corporations overborrowed but were often left with large landbanks bought at the top of the market whose value had declined to less than the value they had used as security for the initial loans. Bad debts in this sector added to the problems of the region's banks. Chinese banks found themselves carrying very large ratios of non-performing loans, potential bad debts, while the assets used as collateral for the loans were now worth less than the debt in a market with few buyers. Added problems for the Chinese banks were the interbank loans they has made to state banks, a particular problem in Indonesia, and loans they had made to politically connected corporations, a particular problem in Thailand. These loans figured very prominently in the non-performing loan portfolios of Chinese banks in Indonesia and Thailand.

The crises faced by Chinese corporations in the regions, however, leads us to one of the methodological problems encountered in the literature on Chinese business: a tendency to both analyse it on a country by country basis and an assumption that the reasons for its success was the close links, 'crony links', of many of the largest corporations and entrepreneurs with their respective governments and state sectors. The crisis has exposed the weakness, once again we might add, of this kind of analysis. What the crisis has shown is that in some cases, corporations that got into serious difficulties in one economy, were both saved by and able to continue successfully in other locations: diversification among economies and regionally-based strategies paid off by spreading the risk.

A clear example of this is the Salim organisation in Indonesia, the group built and controlled by Liem Sioe-leong, reputedly the richest man in Southeast Asia prior to the crisis. Its business activities in Indonesia were seriously affected by the crisis and, as one of the leading 'cronies', Liem was the subject of racist attacks on his Jakarta home at the height of the crisis in April/May 1998, which forced him to flee to the United States. Unable to repay his loans, Liem was forced to surrender control of Bank Central Asia, one of the group's flagships, and pledge shares in his companies to the government restructuring agencies, to the extent of virtually liquidating his equity. While his Indonesian companies were on their knees, however his holding company in Hong Kong, First Pacific Group, was by 1999 making major acquisitions around the region[12] and was able to buy back the pledged controlling equity in some of the Indonesian companies[13]. At the same time, its share price in Hong Kong advanced by more than 60% in 1999. Another example is the Malaysian-based Hong Leong Group controlled by Quek Leng-chan. Despite being hit severely by the economic crisis in Malaysia and the political fall-out from Quek's close links with sacked former Deputy Prime Minister Anwar Ibrahim, it consolidated its strength in the banking sector in Hong Kong[14] during the crisis, culminating in Dao Heng Bank becoming one of the 32 constituent stocks of the Hang Seng Index in December 1999.

Chinese business has been subject to the same pressures as other types of business during the crisis. However, its structure and dynamic, enabled it to survive. The tendency for Chinese business to grow by multiplication of enterprises as new entrepreneurs emerge, rather than by consolidation into fewer larger groups as the economy matured, tells us how. Some entrepreneurs, including some of the largest, have gone under but others have emerged to take their place, whether by design or by necessity, as their employers downsized or went under. New companies have been formed and some of these will grow into the giants of the future. This is what happens as a result of major crises. The substantial transformation in the Taiwanese economy, an economy in which small entrepreneurial companies predominate in the more dynamic sectors, including computers, IT and exporting, enables us to understand what is happening and what we need to look for elsewhere. The move into software and internet services by newly formed companies in Hong Kong[15] and Singapore and even Thailand and Malaysia[16] tells us that these changes are not confined to Taiwan. A new sort of economy is emerging.

Throughout the region, those economies, sectors and entrepreneurs that were not overburdened with the heavy capital investment and infrastructure that was required for Fordist industry have had the opportunity to make the changes required to move into the new growth sectors. In this respect, it is those societies in which entrepreneurial activity has traditionally flourished that have found the energy, initiative and resources to make the change. While the crisis has only endured for less than three years, the fundamental changes in industrial structure, in producing new business leaders and in laying out the path of future development have transformed the region.

Nothing will ever be the same again. East Asia will once again become the engine of growth of the world economy but it will be moving it in a different direction and have a different dynamic. Crisis produces fundamental change forever. The new question will be whether it is becoming increasingly directly competitive with the new American economy and presenting an entirely new type of East Asian challenge to only recently re-established American economic hegemony.

Notes

[1] Other regional stock markets registered the following % gains in 1999: Jakarta 81%, Kuala Lumpur 40%, Tokyo 52%, Seoul 52%, Taipei 31% (*Far Eastern Economic Review*, 13 January 2000, p. 32).

2 Krugman (1998) was fairly typical of this kind of analysis. He has, however, since changed his mind.

3 Radelet and Sachs (1999) come to similar conclusions to this paper in respect of the most important cause of the crash, panic in financial markets.

4 This saga and others can be followed in the pages of the *South China Morning Post* throughout December 1996.

5 I was personally quoted rents of US$10,000 per month for residential accommodation in both Hong Kong and Singapore in January 1997. This was far beyond the capacity of professional incomes in either city.

6 In December 1997, banks from France, the UK and Germany had a combined exposure of US$58.2 billion in South Korea, Thailand, Indonesia, Malaysia and the Philippines and a further exposure of US$120.6 billion in Hong Kong, Singapore and Taiwan. The majority of these banks were not traditional lenders into the region. French banks, Societe Generale, Credit Lyonnais and BNP were owed US$51.2 billion; UK banks, Natwest, Royal Bank of Scotland, Barclays and Schroders US$55.7; and German banks, Dresdner, Bayern Landsbank, Commerzbank, Westdeutsche Landsbank, Deutsche Bank and Bayerische Vereinsbank US$71.8 billion (*Le Figaro, La Vie Economique*, 26 December 1997).

7 Notice the increasing promotion of loans over attracting deposits in bank advertising.

8 Turnover on the Hong Kong Stock Exchange, for example, which had averaged around HK$25 billion per day in the first half of 1997, was in the range of HK$5-8 billion for much of 1998.

9 Two of the government's choices for core banks, Bank Bumiputra and Multi-Purpose Bank, had very poor records of financial management and had been insolvent and required government bailouts in the past. They were, however, in both cases very close to government and the ruling party UMNO.

10 The writer has located a large concentration of Taiwanese IT companies in Dongguan in Guangdong Province that have been established in the last three years. Both R&D and production facilities have been established there.

11 Exceptions to this are the large cigarette manufacturers controlled by the Halim, Wonowidjojo and Sampoerna families, who seem to have come out of the crisis quite well presumably from increased demand for tobacco during the crisis and the industries assured cash flow.

12 First Pacific's most important post-crisis acquisition was Philippines Long Distance Telephone, which when added to the group's already strong position in mobile phones, made it the largest fixed-line and mobile telephone operator in the Philippines.

13 For example, First Pacific is now the largest shareholder in Indofoods, the largest manufacturer of noodles in Indonesia.

14 Hong Leong group own two fully-operational banks in Hong Kong, Dao Heng and Overseas Trust, whose more than 60 branches make it the fifth largest banking group there (after HSBC Group, Standard Chartered, Bank of China group and Bank of East Asia).

15 DHL/SCMP Hong Kong Business Awards 1999 were dominated by Hi-tech companies for the first time (*South China Morning Post*, 10 December 1999, Business p. 7).

16 See the article by Chen May-yee, *The Asian Wall Street Journal*, 17 November 1999, pp. 1, 9, for evidence of these developments in Malaysia and the growing links of these new entrepreneurs with venture capital providers in Singapore.

References

Asian Wall Street Journal, various dates.

Far Eastern Economic Review, various dates.

Institute of International Finance (1999), *Net Private Flows to Emerging Markets*.

Kindleberger, C.P. (1973), *The World in Depression 1929-39*, University of California Press, Berkeley.

Kindleberger, C.P. (1978), *Manias, Crashes and Panics*, Macmillan, Houndmills.

Krugman, P. (1998), *What Happened to Asia?* //www.stern.nyu.edu/~nroubini/ AsiaHomepage. html. 18 January.

Le Figaro,(1997), 'La Vie Economique', 26 December.

Radelet, S. and Sachs, J. (1999), 'What Have We Learned, so far, From the Asian Financial Crisis?' USAID (Contract PCE-Q-00-95-00016-00 Delivery Order 16), 4 January 1999.

South China Morning Post (various dates).

4   Mismatch at the Interface: Asian Capitalisms and the Crisis

CONSTANCE LEVER-TRACY

Introduction

The Asian crisis has been called the 'first crisis of globalisation' (Higgott, 1998, p. 2). Whether or not precursers, in the early 1990s, in Latin America or Scandinavia, might claim that title (Bell and Beeson, 1998, p. 6), it surely manifested itself as the first global crisis of globalisation, leaping from country to country across the region and thence via Russia to Brazil, even at moments threatening Western economies. Its novelty was demonstrated in the fact that it was unpredicted, at least in its timing, the speed of its contagion, its depth and its duration, by any of the models and theories which had been put forward to explain the prior 'economic miracle' in the region and that for nearly two years, varying prescriptions for recovery failed to demonstrate their appropriateness (Krugman, 1998b, p. 42; Sachs, 1997). When the recovery began to assert itself, by mid 1999, it vindicated neither those who expected only a brief temporary interrruption to the upward curve, nor those who rushed to pronounce dead the 'miracle' and any imminent 'Pacific Century'.

Globalisation is widely defined in terms of an exponential increase in transnational flows of goods, information, people and capital, facilitated both by technology and by deregulation[1].The Asian 'economic miracle' was driven by rising manufacturing exports and its financial crisis was preceded by a very rapid growth of massive inflows of short term capital, and was sparked by capital and currency flight of even greater proportions. BRIE argues further that:

> Globalisation, by contrast to internationalisation and multinationalisation should be characterised as an era of multiple methods which originate in a variety of places in the world... as each place – locale, nation or region – builds its response, the variety and multiplicity... is reinforced by

66

globalisation itself. Globalisation then is characterised by heightened uncertainty, by great market volatility... Competition becomes multidimensional as well as multidirectional...(BRIE, 1991, p. 3).

This chapter will take such a world, where there are increasing flows between multiple and diverse centres, as the novel context for the crisis. It will suggest that problems of articulation, between different ways of integrating capitalist operations, are one of its crucial, but so far insufficiently conceptualised and explored, aspects. While concrete individuals, groups or countries are nearly always involved with several forms of integration, there is a tendency for one to predominate in a particular group or place and to set the parameters for the others. Different ways of relating different units of capital have different requirements for effective functioning, different strengths and weaknesses and different paths of development and are likely to degenerate or collapse in different circumstances. Although co-existence and compromise between them is ubiquitous, synergy is unstable. Where the balance between them changes suddenly, a disjuncture at the interface can irrupt into an escalating crisis.

We will focus on three distinctive ideal types of integration, with different relative importance for different kinds of capitalism: hierarchical, free market and horizontally networked. In the first type, integration is through a hierarchical plan transmitted downwards from a centre[2]. The relations between contractors and dependent subcontractors and between a parent company and its subsidiaries could come under this category, but in this paper we will concentrate in particular on the manifestation of the hierarchical plan in the activities and impact of developmental states, through their selective sponsoring, direction and coordination of private capital for collective goals, and through the relationships entered into by corporatised state enterprises.

The second type of integration is through free market relations, composed of independent actors (who may be unknown to each other) entering into sequential, short term contracts for profit. We will be particularly concerned with the extreme manifestation of this kind of relationship in rapidly turning over credit, equity and currency markets. The third form involves integration through long term horizontal networks between corporate bodies (including their continuing association in cartels and *keiretsu*), social collectivities or persons. The main focus in this paper will be on the personalised networks that are embedded in relations of reciprocity based on kinship, friendship or other long term trusting relationships.

Much of the debate between economists and political economists about the credit for the Asian 'economic miracle', about blame for the current crisis or about the way to recovery has focussed on states and markets. The contribution of horizontal networks, has often been completely ignored. This paper will seek to redress some of this neglect. The debate has also tended to be posed in partisan fashion - for one side, against the other - rather than focussing, as we will seek to do, on the opportunities and dangers that result from the interrelationship of the three types of integration.

The aim of the paper is to set out this triple typology and to point to the precarious balance between the three types of integration. It suggests that exploring the recent changes in this balance, and the resulting disjuncture, might be useful in conceptualising and exploring the causes and outcomes of the Asian crisis, although it is beyond the scope of this paper to analyse these in any detail.

Three Kinds of Capitalist Integration

Debate about the key attributes of a so called 'Asian capitalism' can conceal the complexity of the quite different kinds of capitalism currently coexisting and interacting in the region (and within each country). These have made distinctive inputs into development, changing in relative importance over time and, this paper argues, it is the clash at the intersection of their different trajectories that has irrupted into crisis. Still, with considerable simplification, one can propose three types of capitalist development that have been operating - one integrated primarily by the hierarchical plan of developmental states (sometimes characterised as the Japan model), one where free market relations more often connect different agents (a type seen as uniquely privileged in the ideology of Western capitalism in recent times) and one where the preferred relationships involve horizontal networks. In Chinese diaspora capitalism these latter generally take the form of personalised ties between owners, which can often have a transnational dimension, creating links around the region or even the globe (EAAU, 1995)[3].

It should be noted that all three, in varying proportions, can be found in all economies of the region (and no doubt elsewhere), that to varying degrees they cross the boundaries between public and private sectors and that none are uniquely Asian or Western in character.

Nonetheless, different kinds of capitalism give different relative weight to these three forms of integration, combinations that become expected and institutionalised and whose articulation can be disrupted by a sudden change in the balance.

The model of the developmental state involves a strong and unified state, able to plan and carry out a collective purpose, with a degree of relative autonomy or at least strong bargaining power, vis a vis domestic interest groups and foreign capital, which initiates and coordinates the process of economic development. It may centralise and allocate domestic savings or foreign loans, making rapid growth (even without foreign ownership) possible, by guaranteeing levels of debt much higher than any market would consider prudent, or it may assign preferential or monopoly rights or major projects to selected private or public sector firms seen as competent and willing to further its goals. This is done in accordance with an industrial strategy aiming at economic development and modernisation, or at strengthening of the military capacity of the state or at social goals such as changing the relative status of different ethnic groups etc (Wade and Veneroso, 1998; Khoo, 1998). This model thus involves strong and continuing relationships and communication between state agencies and public and private corporations.

Developmental state strategies may be subject to the lure of grandiose projects and to the dangers of corruption and cronyism[4]. It has been argued that the goals of the corporate actors in this model of capitalism tend often to focus on market share and growth in size of units, and that it fitted well with a period when Fordist mass production was the cutting edge of economic power in the world. A long term decline in the efficiency and profitability of this kind of integration has been setting in.

Such a policy driven model and its hierarchical plans was initially central to modernisation and the economic miracle in Korea, as earlier in Japan. In both Taiwan and Mainland China under reform (but not in Hong Kong) the model describes the state direction of resources to what is still an important (often wholly or partly state owned) economic sector, that provides infrastructure and intermediate goods for other economic actors (including smaller private firms, local collectives and joint ventures) (Cotton, 1998; Evans, 1987). In Malaysia, it has been part of the government's NEP strategy to develop a *pribumi* capitalist class, including state technocrats and the nursing of new private operators, who often originate from privatisations of previously state-owned corporations, at preferential prices. In Indonesia, the state drew selected ethnic Chinese tycoons into an alliance with the public sector and military-controlled enterprises and the president's own circle, and allocated to them major projects and monopolies (Jesudasen, 1989; Robison and Rosser, 1998).

Yet all the success stories of developing Asia (even mainland China) have also involved an export orientation for newly established manufacturing companies, involving global markets, beyond the reach of state policy directives[5]. Here another kind of capitalist integration becomes important, involving free market transactions, based on anonymous considerations of individual short term costs and benefits. Its proponents claim that its guiding objective of profit maximisation leads to greatest efficiency. Much of Asian manufacturing production is linked to its major consumer markets, in the West, mainly through open market relations. Furthermore, increasingly and with rapid acceleration in the 1990s, global flows of short term loan capital and equity and currency transactions have multiplied, as competitive deregulation and liberalisation and the technology of the communications revolution have swept round the world and as stock market flotation has taken off.

American interests and global regimes and international treaties, as well as the conditionalities imposed by bodies such as the IMF, the World Bank and the World Trade Organisation, and the influence, for example, of ratings agencies such as Moody's, have pushed outwards the boundaries of this model, against any opposition (Jayasuriya, 1998; Wade and Veneroso, 1998). This free market system is, however, also liable to degenerate and betray its own aims. The dangers of irrational herding, where ignorant investors chase the expected behaviour of other investors, and of self reinforcing upward and downward swings, in a system of multiple, uncoordinated short term decisions, has always been a threat, and obviously remains so. (Sachs, 1997, 1998; Krugman, 1997; Krugman, n.d.).

Both these kinds of integration, in varying combination, are to be found in many parts of the world without producing sustained growth and development. Those who argue their relative merits rarely discuss a third kind that is particularly prominent in Asia, which may well have been the main motor of this distinctive trajectory, especially since the 1980s - the horizontal networks of mainly small and medium, entrepreneurial family businesses, linked transnationally through long term personal relationships of reputation based trust, which characterise, in particular, Chinese diaspora capitalism[6]. As the oldest of the forms, with a history that long predates modern capitalism, its supercession or decline have often been prematurely predicted. However, while some Chinese businesses have established privileged relationships with states and many have increasingly sought to tap into anonymous share and money markets, the large majority of small and medium Chinese businesses have continued to rely heavily (although not exclusively) on network relationships and these remain an important resource for the largest tycoons as well (Lever-Tracy et al, 1996, Ch. 7; Olds and Yeung, 1998; EAAU, 1995).

This Chinese diaspora capitalism is a distinctive type of capitalism. Continuing control by members of entrepreneurial families, a preference for personalised, longstanding, external networks based on trust and often leading to friendship and a strategy of multiplication and diversification are the three legs on which Chinese capitalism has rested in the past and the present. These attributes have survived and flourished despite stock exchange flotation, professional management, Western education, and the attenuation of patriarchy and of regional and dialect divisions. An unusual degree of similarity and continuity between large and small firms has facilitated rapid social mobility and given each some of the strengths of the other, with small firms able to engage in global operations through transnational networks and those that have grown large still enjoying the flexibility derived from personal connections and family control.

These distinctive aspects of Chinese business practice are interrelated and tend to reinforce each other. Friendship between owners with decision making power has both more legitimacy and more reliability than that between possibly impermanent managers subject to bureaucratic norms. Diversification allows for growth and modernisation, while perpetuating a competitive entrepreneurial familism. The flexibility and viability of small firms, motivated by not always unrealistic ambitions to grow and increase their independence, remain a generator of growth and development.

While not immune to motives of growth and profit, this kind of capitalism gives a high priority to independence and the retention of family control. Even the most successful tend to opt for a conglomerate structure of multiple small units, in which professional managers can be coralled and subject to the 'custodial control' of family members (Limlingan, 1986), rather than for the centralised bureaucratic structure that characterises Western and Japanese corporations, and they are always subject to fission, as managers and heirs split off to carve out their own independent paths (Tam, 1990). Unlike a purely free market model, although actively engaged in market competition, this kind of capitalism also remains deeply embedded in social structures of kinship, friendship and community. Unlike the hierarchical model, it is individualistic, with personalised loyalties and limited respect for formal rules and external authority. As a historically constituted diaspora, its networks are transnational and it thus has both local social roots, that are absent in pure market relations, and a global competence, that evades the developmental state (Redding, 1990; Wong, 1988; Hsing, 1998; Lever-Tracy et al., 1996).

Its weaknesses of lack of state support and limited resources, create dangerous temptations to venture into other modes - for a few to establish crony ties to powerful individuals in the state and, for more substantial numbers, to engage in market speculation[7]. This kind of capitalism has always been the dominant form in Hong Kong, and has become such in Taiwan. In Southeast Asia, as also among Chinese business communities in North America and Australasia and elsewhere, it represents the most common form of Chinese economic activity (Omohundro, 1981; Lever-Tracy et al. 1991; Hamilton, 1991; Mackie, 1992; Hsing, 1998; Light, 1972; Wong, 1988).

Table 4.1 summarises the distinctive goals, needs, limits, dangers and tendencies of these three types. Their different strengths and weaknesses may lead to advantageous complementarity and cooperation. The Asian economic 'miracle' was a product of such a synergy, manifested in the effectiveness of a competitive assault on global markets, coordinated by developmental states and the institutional linkages of Japanese *keiretsu* and the personalised networks of diaspora Chinese capitalism[8]. Yet it is also clear that the complementarity may break down and the needs of one system may represent dangers for another.

Explaining The Asian Crisis

Through much of the 1990s the Japanese economy stagnated. In January, 1997 Hanbo Steel, a large Korean chaebol, collapsed under $6bn in debts - the first bankruptcy of a leading Korean conglomarate in a decade. In mid May, Thailand's baht currency was hit by a massive attack by speculators. Thereafter the crisis deepened and spread rapidly throughout the region. Only by early 1999, were there at last indications of an end to the slide and some projections of slow recovery, although these did not seem to be clearly related to the medicines prescribed.

What started as a financial and currency crisis devastated the real economy of much of the region, as the costs of imported food, of necessary inputs for production and of debt servicing and repayment all soared.

Table 4.1 Three Types of Capitalist Integration

	Hierarchical Plan	Free Market Relations	Horizontal Networks
Goals	Long term national or social goals. Rapid growth. Market share.	Short term individual profit maximisation.	Large ambitions. Family control. Multiplication of small units. Diversification.
Needs	United purpose. Relative autonomy to implement goals. Legitimacy and loyalty. Official secrecy.	Rule of law to ensure level playing field. Transparency to enable rational decisions by unrelated agents.	Personal trust and reputation. Confidentiality.
Strengths	Concentration of resources. Large scale. State support.	Global scope. Competitive efficiency.	Entrepreneurship. Flexibility. Transnational networks. Local knowledge. Long term relationships.

Weakness	Effectiveness restricted within national borders. Red tape.	Short term perspective. Lack of local knowledge.	Small scale. Lacks state support.
Dangers	Corruption, cronyism, grandiose projects leading to loss of collective purpose or autonomy or legitimacy.	Irrational herding. Bubbles and crises.	Temptation of a) state cronyism (may rebound) or b) free market speculative activities.
Tendency	Suits period of Fordist mass production and national economies. Recent tendency to stagnation or decline.	Increasing due to US pressure, deregulation, global regimes (IMF, WTO, ratings agencies etc), information technology.	Growing importance in era of global flexible economic activity.

New credit for productive activities dried up or became phenomenally expensive and regional markets shrivelled while cut throat competition for exports grew and country after country (with the notable exception of both Taiwan and mainland China) experienced negative or near zero growth.

Policy driven state and institutional relationships, market forces and social networks exist, as we have said, in varying combinations in all capitalist systems. Yet even before the current Asian crisis threw all prior trajectories and projections into doubt, the continuing coexistence of these three models was frequently questioned, especially by monist partisans of states or markets. The futuristic triumphalism of the advocates of Japanese forms (for Western as well as third world countries) became more muted over recent years, as the claims of convergence to the market model grew louder, decrying the other two as sub optimal relics or at best transitional forms soon to be phased out[9].

While the miracle seemed set to continue, proponents of developmental states and of free markets were each happy to assert the predominance of their chosen model and to claim for it the credit. Now they each began to blame the other, attributing responsibility either to state intervention or corruption or to global capital market speculation[10]. There is, however, no satisfactory explanation of why a previously supposedly dominant and successful mode should so suddenly have been displaced by its rival. Slightly more sophisticated versions argue not displacement at the helm, but rather a process of decline or degeneration of an optimal mode, with states deprived of effectiveness or autonomy by global market forces or else with markets corrupted by increasing state intervention. This paper will suggest rather that a changing balance, between different kinds of economic logic, have irrupted into a clash and mutual undermining at the interface between them. The progressive deregulation of global financial markets, and consequent massive increase in footloose capital flows (Helleiner, 1995), resulted in an escalating incompatibility that ceased to be manageable.

IMF spokespeople and a variety of economic commentators have speedily vied to define the sins of Asian regimes that have brought on such an outcome. But as Janet Yellen (1998) has pointed out,

> Ex post attempts to identify the fundamental causes of a financial crisis always suffer from the problem of distinguishing insight from hindsight. Any financial journalist today can tell you that the crisis was the inevitable consequence of overvalued exchange rates, large current account deficits, short-term capital inflows, opaque financial systems, 'crony capitalism,' more generally the problems of the 'Japanese model' of capital markets, or one of several other supposedly fatal flaws in East Asian capitalism. But it seems fair to say that a year ago nobody suspected that such a calamity was remotely possible, although all of what are now described as the fatal flaws of the East Asian economies were reasonably widely understood even then, at least by experts.

Clearly these 'flaws' had not prevented the accomplishments of the previous years and indeed, as Joseph Stiglitz of the World Bank noted (1998b):

Many of the factors identified as contributors to East Asian economies' current problems are strikingly similar to the explanations previously put forward for their success. Addressing information problems in an effective way, including through business-government coordination, was considered a hallmark of these economies' success; but this coordination is now viewed as political cronyism and lack of transparency is viewed as one of the main failings. Openness to international markets was hailed as one of the grounds of their success, yet insistence on eliminating barriers in capital and trade flows is an important ingredient in many of the reform packages... Promoting competition, especially through export-oriented policies, was hailed as one of the central pillars of their stellar performance, yet lack of competition in the business conglomerates is seen as one of the critical failings. Finally... what were previously viewed as strong financial markets, which were able to mobilize huge flows of savings and allocate them remarkably efficiently, have turned into weak financial markets which are blamed for the current crisis.

In general, he pointed out, most of the macroeconomic fundamentals, so dear to economists, had previously been considered sound. These countries had high saving rates, government surpluses or small deficits, low inflation, and low levels of external debt relative to other developing regions. Explanations in terms of retribution for Asian sins is the less convincing in that the crisis has affected in quick succession economies operating in quite different ways and with quite different circumstances and weaknesses. For example, although some countries (Thailand) had relatively large current account deficits, other crisis counties had relatively standard current account deficits (Indonesia), and others were not only moderate but falling (Korea and Malaysia).

Conversely, some analysts have suggested that it is not impossible that the entire explanation for the crisis could lie in market forces run wild, a crisis set off by a self perpetuating vicious circle of panic, among fund managers and speculators who control vast, short term capital flows. Specific local or temporary problems could have sparked a capital flight, which then fed on itself and became a self fulfilling prophecy of doom for economies (and their currencies) that had allowed themselves to become too dependent on such loans (Sachs, 1998 ; Stiglitz, 1998a; Shiller, 1989). Paul Krugman has given a terrifying description of the ignorance, short term thinking, greed and herding instincts evinced at a meeting of money managers, who between them controlled the placement of billions of dollars of other people's money to emerging markets of which they knew little and understood less (n.d.)[11].

With more emphasis on the ways their champions may themselves have been subverted by their rivals, some statists have interpreted with alarm the growing power of untramelled markets to weaken, divide and disarm developmental states, thus laying them open to the loss of their relative autonomy and thus to corruption and capture by sectional interest groups. (Kim, 1998; Moon and Rhu, 1998; Robison and Rosser, 1998). As collective purpose and competence faded, states served rather to guarantee excessive debt for wasteful over investment in gradiose projects, for towers and bridges and national car industries, and continued to prop up inefficient state owned sectors, placing eventually intolerable burdens on their banking systems.

Alternatively, what are seen by market liberals as the opaque and selective machinations of corrupt states are claimed to have introduced moral hazard and a liability to irrational panic into the operation of markets, preventing them from producing the expected prudent and rational outcomes. The privileged personal or institutional relationships, the secretiveness and the particularistic trust that characterise both developmental states and networks are seen as undermining the system trust on which markets rely. In the absence of proper transparency and prudential governance, rational market decision making becomes herd instinct (Krugman, 1998a).

Mismatch at the Interface

None of these explanations, by themselves, provide a fully satisfactory answer to the question of why the capital flight spread so rapidly and extensively through and across the economies of Asia in particular. This is particularly the case when we consider the speed of the contagion, which extended the crash from firms and banks that were shaky, to the majority that were initially sound, and from one country to other unrelated and dissimilar ones, throughout the region. Our argument is rather that we must also focus on a clash at the interface of potentially incompatible systems, which had been previously developing along partially independent trajectories, due to a sudden change in the balance between them.

One of the most obvious new factors in the situation was the rising tide of volatile, short term hot money flowing around the world, at ever greater speed, since deregulation and the big bang of 1986, when world money markets were linked up by computer, producing enormous vulnerability. While some Asian economies (but not others) have always had a ratio of debt to equity that would be considered unacceptable in Western capitalism[12], the relative scale of short term loans, and the focus on recycling short term debt, to cover long term projects, has certainly grown massively in recent years.

Indonesia, Malaysia, South Korea, Thailand and the Philippines, had net private inflows of foreign capital of $41bn in 1994 jumping to $93bn by 1996. Much of the rise was in commercial bank loans, which shot up from $24bn in 1994 to $56bn in 1996. When the crisis hit, it was such short term bank loans that fled, swinging by some $70bn to net repayment of $21bn in 1997, out of a total capital swing of $105bn, equal to 10% of the GNP of these countries (Wolf, 1998). Portfolio capital increased from 2% of net capital flows to developing countries in 1987 to 50% by 1996. (Winters, 1998, p. 3).

These flows came into an environment where both networks and developmental states were unprepared to evaluate the risks of such borrowing, and where the lenders of the new hot money were similarly ill equipped to understand the local mechanisms for assessing lender risk. Almost all analysts, irrespective of their differing interpretations, agree that a common lack of general public transparency, associated with both hierarchical plans and horizontal networks, posed major problems for outside investors, particularly those involved in making rapid short term decisions on the assumptions of free market relations. Risk assessment was made harder by lack of reliable information about the level of indebtedness and likely proportion of bad debts of banks or about the real value, profitability and ownership structures of companies and of the assets that stood as security for loans. Adherence to prudential rules and accounting standards could not be relied on and many of the formal oversight procedures that Western capitalism has developed to guarantee system trust were inadequate[13]. When doubts about particular transactions raised more general suspicions, there seemed to be no reliable way of distingusihing between more or less sound projects, institutions or economies. Western ignorance combined with Asian opacity to produce an undiscriminating perception of risk throughout an undifferentiated 'Asia'. The clearest vindication of this is the speed with which a relatively minor crisis in Thailand became a regional crisis as international banks called in loans and portfolio managers sold down their holdings of stocks throughout the region.

As we have seen in the previous section, the various kinds of capitalism in Asia are not lacking in means to assess risk and establish trust. These mechanisms, involving long term vertical or horizontal relationships between persons, networks or institutions and depending on the indispensability of reputation for future access to opportunities, are very different from those that have become central to the operation of short term investors and lenders on global financial markets[14]. This is not primarily a problem of excessive weakness, incompetence or corruption (all of which are to be found in good measure in all societies) but of a mismatch, a lack of synergy, at certain interfaces between different kinds of capitalist integration. Both formal rules and long term relationships exist, of course, in all modern economies. The balance between them differs, however. All economies are part of an interconnected global system and engage in continuing relationships of trade, investment, subcontracting etc. However the particular interface, between different currents in the global ocean, that is represented by the short term money market, is one where the misalignment is at its most extreme and seems to have become catastrophic.

Such a mismatch has also affected those whose primary integration is through networks. For Chinese diaspora capitalists the temptation has always been for some of them to allow themselves to be sucked into the circuits of the other two, seemingly more powerful and wealthy, systems. With their business competence and regional or global ties, a small number of Chinese tycoons were able to bargain for themselves privileged connections with developmental states or with sections of a fragmenting political or military elite, even where there has been discrimination against the ethnic group as a whole. The dangers of being tarred with the unpopularity of corrupt regimes in situations of economic or political crisis, and of transmitting this to the whole Chinese community, have become only too visible in Indonesia, when they were targeted by rioters. It is clear too that the whole of that country suffered from the alienation of its most dynamic capitalist group and that as a result, its economic recovery was postponed. The appeal of global markets, on the other hand, and heavy involvement in property and stock market speculation, led to major losses and collapses in Hong Kong and elsewhere.

Thus the incompatibility of different kinds of capitalist integration has irrupted in a clash at the interface. The liberalisation and deregulation of global capital markets have undermined the integrity of developmental states (Robison and Rosser, 1998, p. 12)[15]. The corruption of these, in turn, has accentuated 'moral hazard' for market participants, creating conditions for irrational bubbles and disproportionate panics which, through ignorantly spread contagion, undermine even the healthy parts of the real economy. Some sections of the networked diaspora are drawn into global speculation or crony capitalism, paying a price in loss of control and reputation and attracting scapegoating attacks on the whole community. In turn, through their particularistic relationships they contribute to the undermining of systemic trust in both the level playing field of the market and the legitimacy of the state's public purpose.

Remedies and Prospects

In this situation of escalating mutual undermining at the interface of the different systems, proposed remedies have ranged from a degree of insulation to the elimination of alternative forms. Jayasuriya (1998) sees the emergence of pressures to establish an authoritarian liberal global order in which the purpose of the state is to safeguard the freedom of the market and to insulate it from politically or socially motivated interference. 'An independent central bank strikes at the core of the developmental state' (p. 17) as it does at that of the welfare state. Through a new constitutional order of international treaties and institutions, with a strongly juridical cast (competition and anti trust laws, intellectual property laws, credible financial market law) from which none can afford to opt out, the global market would be disembedded and its autonomy established.

The longer term remedies proposed as conditions for IMF bailout funds have tended (in accord with US pressures) to include measures aimed at asserting the primacy of markets and at transforming the capitalism promoted by a developmental state into a form convergent with what is seen as the Western model[16]. Government industry policies, selectivity and guarantees were to be replaced by internationally accepted prudential and transparency standards and by open borders and level playing fields. The system of long term relationships, cross shareholdings and preferential treatment between economic institutions was to be ended, and replaced by competition. Some statements have gone further and aimed at the elimination of the policy directed form altogether. In testimony before the Committee on Agriculture, US House of Representatives (May 21, 1998) Alan Greenspan called for the US Congress to increase support for the IMF saying:

> Some observers have also expressed concern about whether we can be confident that IMF programs for countries, in particular the countries of East Asia, are likely to alter their economies significantly and permanently. My sense is that one consequence of this Asian crisis is an

increasing awareness in the region that market capitalism, as practiced in the West, especially in the United States, is the superior model. I believe IMF personnel are committed to this[17].

The failure of the IMF immediate remedies led, on the contrary, to proposals to manage the clash at the interface from the opposite direction, through government controls at least over international short term capital and currency flows. While the policies of Mahathir were the first concrete implementation, the approach had appeal in many Asian quarters and, with the spread of the crisis, amongst some economists, and he found unexpected support even from George Soros (*FEER*, 16 July, 1998, p. 86). Even *laissez-faire* Hong Kong was forced to intervene massively on two occasions to defeat speculative attacks on its currency and stock market. The name of Keynes began to appear with unwonted frequency in unlikely places such as the pages of the *Far Eastern Economic Review*.

Neither approach has been entirely convincing or consistently followed. The benefits of world trade over a prolonged period, the massive failures of autarkic regimes in places like Burma or North Korea, and the irreversible globalisation produced by technology, make it very unlikely that any country will seek to cut itself off from global markets. While short term flows were reversed during the crisis, substantial long term FDI investment continued to flow in, if more slowly, to all countries except Indonesia (*FEER*, 8 April, 1999, p. 62). On the other hand, the implications of wholesale dismantling of previously successful forms of capitalist integration has led to hesitation amongst its advocates and resistance by the recipients. It is significant that both Mahatir and the IMF have retreated from the full implications of their own policies, the more so as neither approach has demonstrated any clear superiority over the other. In March 1999 there was little to choose between the faltering bottoming out in Malaysia and in Thailand (*FEER*, 24 September 1998, p. 18; 28 January 1999, pp. 42-44 and 55; 18 March 1999, p. 61). At the time of writing, the continuing American boom has restored the faith of many in free market liberalism, but any future downswing in the roller coaster is likely to garner more supporters for attempts to establish cooperative international or regional political control over global financial markets. A degree of insulation at this perilous interface including, on the one hand, restrictions on short term foreign borrowing and, on the other, much stricter requirements for transparency where it does occur, seem to be a necessary minimum to attenuate future crises.

Chinese Network Capitalism - Surviving the Crisis

Network capitalism may have some advantages over the other two kinds in resolving the dilemmas of such a clash at the interface. While it may have little chance of dominating the others, its practitioners have a long history of successful evasion of state controls and of establishing personal trust and reliable sources of inside information in what are supposed to be anonymous markets.

> With a thousand years or more of continuous experience of dealing with states and markets, East Asian social networks are pre-adapted to coping with these rival political-economic forms, and quite effective at prevailing against them (Winckler, 1988, p. 280).

A family fortune, structured as a loose congeries of small firms, designed to be split up amongst the heirs, can relatively easily scale down its operations or retreat from parts of them. In a society with a constant turnover of small firms, individuals are accustomed to starting again and to moving in and out of paid work and of different business ventures. They should often be able extricate themselves from their recent, over extended involvement with global money and equity markets and return to more traditional network sources, where personal trust can be maintained when market trust has collapsed. Unlike the capitalism of developmental states they have the resources to sustain an involvement in regional or global activities through their transnational networks, even if they are shut out of (or cut themselves off from involvement in) anonymous global market relations.

Just as so many academic observers have been able to discuss the 'Asian miracle' without noticing the key role of the transnational network capitalism of the Chinese diaspora, so they discuss strategies for recovery from the crisis that ask no questions about its possible contribution. One might hypothesise that the capacity to survive, or to retreat and later to revive of those more rooted in transnational networks may be more substantial than has been recognised. Those who are ensconced in such networks may sometimes establish privileged relationships with states and may venture successfully into anonymous global markets and short term speculation, but are they not, perhaps, equipped to continue operating if these fail, relying on social relations and informal sanctions to uphold trust, and on their own linkages to operate across national borders?

Notes

[1] Some world systems theorists question the extent and unprecedented significance of such increases, in the perspective of the *longue duree*. However, the data on many transnational flows certainly indicate a recently accelerating renewal of processes that were stalled or even reversed in the first half of the twentieth century. Global trade did not recover its level, relative to global production, of 1914 until 1980 (Schwartz, 1994, p. 4). Between 1959-1972 average annual imports constituted less than 9% of the GNP of the OECD countries. For 1972-1987 the proportion had risen to over 15% (ibid p. 283). Between 1980 and 1991 the stock of international bank lending, as a proportion of the GDP of these countries rose from 4% to 44% and that of internationally issued bonds rose from 3% to 10% (ibid, p. 238). See also Lash and Urry (1987), Ch. 1. While there have been periodic crests and troughs of global contacts through human history, the cumulative nature of advances in the technology of transport and communication means that the current period cannot be simply interpreted as 'back to the future'.

[2] Insofar as this typology is intended to refer to the relations between distinct economic actors, it does not include either the dictates of the head offices of corporations or the state directives of autarkic command economies. The plans of pre-reform Russia or China would therefore not have a place in this classification, which is not intended to be exhaustive of all forms. Their exclusion derives from their very limited interaction with other, distinct units of capital.

[3] Small firms in the 'Third Italy' or in Silicon Valley etc may also relate through such horizontal networks. Other diasporas may also establish transnational networks. However, in the context of East Asia and of both the 'Asian Miracle' and the Asian crisis, the importance of the Chinese far outweighs any others. See Lever-Tracy et al, (1996), Chs 2, 16, for a fuller discussion of this.

[4] Similar claims have been made about the 'military industrial complex' in the United States.

[5] Persistence in autarkic or inward looking developmental state policies have had mixed effects in India or Vietnam and absolutely negative ones in Burma and North Korea, and the Asian miracle took off elsewhere only with their abandonment. As well as global markets, foreign capital has often and foreign technology has always played a role in the policies of such developmental states but in no case, except Singapore, have transnational corporations been seen as the primary agents.

[6] Personal networks are, of course, not unique to Chinese business relations, nor are they necessarily ethnically exclusive. Chinese networks, however, have an unusual degree of cultural legitimacy, are facilitated by the size and business focus of the Chinese diaspora and have become more prominent recently as they have facilitated their success as investors in Mainland China (Hamilton (ed.), 1991; Hsing, 1998).

[7] There are strong Chinese cultural predispositions to gambling, which encourage the latter (Oxfeld , 1993, Ch. 4; Ho, 1997).

[8] Evans argued that the low levels of direct Western multinational ownership were crucial to successful Asian development in most of the countries of the region (Evans, 1987).

[9] Advocates of free market relations have often claimed maximum flexibility for their model. This is, however, very questionable. An environment of purely short term commitments and insecurity is a disincentive to risk taking, and leaves few willing to

make resources available for rapid change. Arguably, without either a welfare state or the mutual aid of social networks, market based societies would collapse.

[10] For one example of many see Wolf, 1998 or for one from the opposite perspective see Wade and Veneroso, 1998.

[11] The picture makes very unlikely the suggestion of a super intelligent plot by speculators.

[12] Wade and Veneroso, in 'The Asian Crisis' (1998), a piece with particular application to Korea, see high debt levels as a crucial part of an Asian model in which foreign ownership is avoided and governments, through directing bank loans, are able to plan economic development.

[13] See //www.stern.nyu.edu/ ~nroubini/AsiaHomepage. html passim.

[14] It used to be sometimes supposed that a dependence on particularistic trust would severely restrict the growth and operations of Chinese businesses and that a lack of transparency in family controlled firms would prevent them successfully engaging in stock market flotations. In fact they were engaging in stock market floats with great success prior to the crisis, relying on trust in the personal reputations of the company heads, in a system where the sustaining of such reputations is recognised to be of paramount importance (Lever-Tracy et al, 1996, pp. 30-33).

[15] This should not be seen as simply an attack by Western on Asian capitalism. The largest source of short term loans to Southeast Asia came from Japanese banks, and domestic holders (fearful of inability to repay dollar denominated loans) were prominent in the flight from the currencies (Robison and Rosser, 1998, p. 12).

[16] We should note that capitalism in the West has evolved in this direction only recently.

[17] Reported in //www.stern.nyu.edu/~nroubini/AsiaHomepage.html. The appeal failed. A number of writers, with mixed feelings, believe in the inevitability of a general convergence to a Western, market based model and suggest that 'Asian Capitalism' can be seen as a type whose time has passed or as a historical detour now returning to a unitary mainstream (Cotton, 1998, p. 5; Moon and Rhu, 1998, p. 23).

References.

ARC (Asia Research Centre, Murdoch University) Workshop (1998), *From Miracle to Meltdown: The End of Asian Capitalism,* Fremantle, Western Australia, (August).

Bell, S. and Beeson, M. (1998), 'Australia in the Shadow of the Asian Crisis', ARC Workshop.

BRIE (Berkeley Round Table on the International Economy), (1991), 'Globalisation and Production', *BRIE Working Paper* 45, University of California, Berkeley.

Cotton, J. (1998), 'Singapore, Taiwan and the Asian Financial Crisis: The Profits and Perils of Enterprise Association', ARC Workshop.

EEAU (East Asia Analytic Unit), (1995), *Overseas Chinese Business Networks in Asia,* AGPS, Canberra.

Evans, P. (1987), 'Class, State and Dependence in East Asia: Lessons for Latin Americanists', in Deyo, F. (ed.) *The Political Economy of the New Asian Industrialism,* Cornel University Press, Ithaca.

FEER (Far Eastern Economic Review), various dates.

Hamilton , G. (ed.) (1991), *Business Networks and Economic Development in East and Southeast Asia,* Centre for Asian Studies, University of Hong Kong, Hong Kong.

Helleiner, E. (1995), 'Explaining the Globalisation of Financial Markets: Bringing States Back in', *Review of International Political Economy*, Vol 2, no. 2, Spring.

Higgott, R. (1998), 'The International Relations of the Asian Economic Crisis: A Study of the Politics of Resentment', ARC Workshop.

Ho, R. (1997), *Diaspora Chinese Capitalism: a Chop Suey of Rational and Speculative Ventures*, B.A. Honours Thesis, Sociology, Flinders University of South Australia, Adelaide.

Hsing, Y.T. (1998), *Making Capitalism in China: The Taiwan Connection*, Oxford University Press, New York.

Jayasuriya, K. (1998), 'Authoritarian Liberalism, Governance and the Emergence of the Regulatory State in Post-Crisis East Asia', ARC Workshop.

Jesudason, J.V. (1989), *Ethnicity, the Economy and the State, Chinese Business and Multinationals In Malaysia*, Oxford University Press, Singapore.

Khoo, B.T. (1998), 'Economic Nationalism and its Discontents: Malaysian Political Economy after July 1997', ARC Workshop.

Kim, H.R. (1998), 'Fragility or Continuity? Economic Governance of East Asian Capitalism', ARC Workshop.

Krugman , P. (1997), 'Bahtulism: Who Poisoned Asia's Currency Markets?' //www.stern.nyu.edu/~nroubini /AsiaHomepage. html August 14.

Krugman, P. (1998a), 'Will Asia Bounce Back?', speech for Credit Suisse First Boston, Hong Kong, March //www.stern.nyu.edu/~nroubini/ AsiaHomepage. html

Krugman, P. (1998b), 'Saving Asia', *Time,* September 7th.

Krugman, P. (n.d.), 'Seven Habits of Highly Defective Investors', /www.stern.nyu. edu/ ~nroubini/ AsiaHomepage. html.

Lash, L. and Urry, J. (1987), *The End of Organised Capitalism*, University of Wisconsin Press, Madison.

Lever-Tracy, C., Ip, D. and Tracy, N. (1996), *The Chinese Diaspora and Mainland China: An Emerging Economic Synergy*, Macmillan, Houndmills.

Lever-Tracy, C., Ip, D. Kitay, J. Phillips, I. and Tracy, N. (1991), *Asian Entrepreneurs in Australia: Ethnic Small Business in the Chinese and Indian Communities of Brisbane and Sydney*, OMA/AGPS, Canberra.

Light, I. (1972), *Ethnic Enterprise in America: Business and Welfare Among Chinese, Japanese and Blacks*, University of California Press, Berkeley.

Limlingan, V.S. (1986), *The Overseas Chinese in ASEAN: Business Strategies and Management Practices*, Vita Development Corporation, Manila.

Mackie, J. (1992), 'Overseas Chinese Entreprenurship', *Asian-Pacific Economic Literature*, Vol 6, no.1, May.

Moon. C.I. and Rhu, S.Y. (1998), 'The State, Structural Rigidity and the End of Asian Capitalism: a Comparative Study of Japan and South Korea', ARC Workshop.

Olds, K. and Yeung, H. (1998), 'Reshaping Chinese Business Networks in a Globalising Era', *Workshop on Asian Business Networks,* National University of Singapore, Singapore, March.

Omohundro, J.T. (1981), *Chinese Merchant Families in Iloilo. Commerce and Kin in a Central Phillipine City,* Ateneo de Manila University Press, Quezon City.

Oxfeld, E. (1993), *Blood, Sweat and Mahjong: Family and Enterprise in an Overseas Chinese Community,* Cornell University Press, Ithaca.

Redding, S.G. (1990), *The Spirit of Chinese Capitalism*, Walter de Gruyter, Berlin.
Robison, R. and Rosser, A. (1998), 'Surviving the Meltdown: Liberal Reform and Political Oligarchy in Indonesia', ARC Workshop.
Sachs, J. (1997), 'The Wrong Medicine for Asia' reproduced from *The New York Times*, //www.stern.nyu.edu/~nroubini/AsiaHomepage. html
Sachs, J , (1998), 'To Stop the Money Panic, Disclosure is the Key', World Economic Forum Annual Meeting, 13 February.
Schwartz, M., (1994), *States Versus Markets: History, Geography and the Development of the International Political Economy*, St. Martin's Press, New York.
Shiller, R. (1989), *Market Volatility*, MIT Press, Cambridge.
Stiglitz, J. (1998a), 'Bad Private Sector Decisions', *Wall Street Journal*, 4 February.
Stiglitz, J. (1998b), 'Keynote Address' to the Asia Development Forum, Manila, the Philippines, March 12th.//www.stern.nyu.edu/~nroubini/AsiaHomepage. html.
Tam, S. (1990), 'Centrifugal Versus Centripetal Growth Processes: Contrasting ideal Types for Conceptualising the Development Patterns of Chinese and Japanese Firms', in Stewart Clegg and S. Gordon Redding (eds), *Capitalism in Contrasting Cultures*, Walter de Gruyter, Berlin.
Wade, R. , and Veneroso, F. (1998) 'The Asian Crisis: The High Debt Model Versus the Wall Street-Treasury-IMF Complex', *New Left Review*, 228, March/April.
Winckler, E. (1988), 'Globalist, Statist and Network Paradigms in East Asia' in Winckler, E. and Greenhalgh, S. (eds), *Contending Approaches to the Political Economy of Taiwan*, An East Gate Book, Armonk.
Winters, J. (1998), 'The Financial Crisis in Southeast Asia', ARC Workshop.
Wolf, M. (1998), 'Too much Government Control', *Wall Street Journal*, 4 February.
Wong, S.L. (1988) , 'The Applicability of Asian Family Values to Other Sociocultural Settings', in eds. Berger, P. and Hsiao, M.H.H. *In Search of an East Asian Development Model*, Transaction Books, New Brunswick.
Yellen, J. (1998), 'Lessons from the Asian Crisis', address to the Council on Foreign Relations, New York, April 15 //www.stern.nyu.edu/~nroubini/AsiaHomepage. html.

5 Managing Crisis in a Globalising Era: The Case of Chinese Business Firms from Singapore

HENRY WAI-CHUNG YEUNG

Introduction

Since the early 1970s, Chinese business has been serving as a dominant mode of capitalism in Southeast Asia. Events in the second half of 1997 and the whole of 1998, however, have seriously challenged Chinese capitalism in East and Southeast Asia. Ethnic Chinese entrepreneurs have now come to realise the importance of crisis management in a globalising era. Two challenges are particularly relevant to the future of Chinese capitalism in the Asia Pacific.

First, economic globalisation has created tremendous opportunities for Chinese business. The globalisation of Chinese business firms, nevertheless, is a complex process of simultaneously maintaining the core characteristics of Chinese business practice across borders and creating new network competence through strategic enrolment of Chinese **and** non-Chinese actors (Olds, 1998; Olds and Yeung, 1999; Yeung and Olds, 2000). Globalisation essentially means global competition, a problem aggravated by the fact that Chinese business firms are now increasingly competing in foreign lands and/or on unfamiliar turf. The successful negotiation of globalisation challenges by the Chinese business system requires dynamic transformations in its system configurations and institutional contexts (Yeung, 1999a; 1999b).

Second, vulnerability and crisis tendencies are the defining characteristics of economic globalisation because intense cross-border flows of capital and information create a complicated situation of global interdependence and interconnections (Dicken, 1998). As evident in the 1997/1998 Asian economic crisis, a country specific financial crisis has eventually resulted in a region-wide economic crisis and, in some badly affected Asian countries, a political crisis. Insofar as Chinese business is concerned, the challenge here is to manage the crisis through highly proactive and flexible responses to the problems of liquidity squeezes, unrecoverable loans, currency fluctuations, rapid depreciation of assets and general economic recessions.

In this paper, I argue that entrepreneurship plays a critical role in the management of crisis by Chinese business firms from Singapore, many of whom are engaged in transnational operations. When this entrepreneurship is activated in the cross-border operations of Chinese business firms, it becomes 'transnational entrepreneurship' defined as an ongoing process of calculated risk-taking and foresight in foreign business venturing. As many Chinese entrepreneurs today are increasingly operating in different host countries and regions, transnational entrepreneurship is gaining more importance because successful cross-border business ventures require not only the OLI advantages (ownership-specific, location-specific and internalisation advantages) as postulated in Dunning's (1988; 1993) eclectic framework of international business, but also highly proactive and flexible approaches taken by the transnational entrepreneurs themselves.

These attributes of transnational entrepreneurship are useful to our understanding of how some Chinese business firms have negotiated and managed the recent Asian economic crisis. In particular, I examine three strategies of crisis management by leading Chinese business firms from Singapore:

- Geographical diversification of business interests;
- Financial restructuring through tapping into global capital markets and
- Changing modes of foreign market entry.

The data for this paper originate from an ongoing research project in which personal interviews were conducted with top executives from over 200 parent companies in Singapore and over 50 Singaporean entrepreneurs in Hong Kong, China and Malaysia.

This paper is organised into three sections. The next section examines the challenges to Chinese business in a globalising era. I emphasize how transnational entrepreneurship allows Chinese business firms to meet the challenges of globalisation. A brief introduction to the salient features of the recent Asian economic crisis is offered. Section Two is concerned with three strategies of crisis management by Chinese business firms from Singapore. Although empirical data are provided to support my arguments, they are by no means generalisable to Chinese business firms in other contexts. My analysis is therefore best regarded as illustrations rather than universal proofs of the behaviour of Chinese business firms in a globalising era. The concluding section draws some implications for research and policies on Chinese business firms and Chinese capitalism in the Asia Pacific region.

Challenges to Chinese Business in a Globalising Era

In this section, I want to challenge the existing literature on the 'Overseas Chinese'[1] in which Chinese business networks are seen as almost exclusively constituted by actors of ethnic Chinese origin and embedded in an institutional context that is predominantly reflective of **domestic** (i.e. Asia-based) factors, conditions and agendas (e.g. Lim and Gosling, 1983; Redding, 1990; Hamilton, 1991; Brown, 1995; East Asia Analytical Unit, 1995; Fukuyama, 1995; Hodder, 1996; Weidenbaum and Hughes, 1996; Haley et al., 1998; Hefner, 1998). I argue that faced with pressures from the professionalisation of management, global competition and the ongoing Asian economic/ liquidity crisis, Chinese business firms have sought to link up with an increasing number of non-Chinese actors and institutions. Further, these non-Chinese actors and institutions have proactively engaged with Chinese business networks and firms in order to exploit them in a material sense (Olds and Yeung, 1999; see also Lever-Tracy et al., 1996; Hsing, 1998; Smart and Smart, 1998).

I want to emphasize the role of transnational entrepreneurship in enhancing the globalisation of Chinese business firms. The reshaping of Chinese business networks and transnational entrepreneurship are important avenues for us to understand the strategies of crisis management by Chinese business firms from Asia.

Globalising Chinese Business Firms and Transnational Entrepreneurship

International business has become one of the most important fields of Chinese business activity in today's globalising world economy. To strengthen their competitiveness, some Chinese business firms from Southeast Asia have internationalised into new markets and business fields (see Brown, 1998; Pananond and Zeithaml, 1998). Other Chinese business firms have also actively sought new management and financial resources. The change in the global business environment, for example, has rendered market conditions more competitive and shifted management skills away from those developed by their founders. When Chinese business firms extend their operations across borders to become transnational corporations (TNCs), they are often entering into host business environments which are fundamentally **different** from their 'home' countries[2] in terms of institutional and market structures, industrial organisation, social relations and cultural practices. To overcome these barriers to globalisation, ethnic Chinese TNCs need actors who are creative, proactive, adaptive and resourceful in different countries; these are all aspects of transnational entrepreneurship. These actors are often the owners or founding entrepreneurs, who participate actively in the establishment and management of foreign operations. An understanding of the nature, *modus operandi*, performance of these entrepreneurs is vital to the success of international operations by these Chinese business firms.

Despite decades of entrepreneurship research since Joseph Schumpeter and others, however, we know very little about the real **actors** and their behaviour in transnational corporations. In the case of Chinese business literature, this lacuna is attributed to the fact that most studies of Chinese entrepreneurship tend to focus on ethnic Chinese in their **domestic** setting. These studies are concerned with the role of Chinese entrepreneurs in innovation, new business start-ups and economic development of their 'home' countries in East and Southeast Asia. Moreover, Chinese entrepreneurship research has little interaction with mainstream research on international business and organisational behaviour. The latter is preoccupied with the firm as their central unit of analysis. While Chinese entrepreneurship research tends to ignore entrepreneurs in international business, studies of international business and organisational behaviour focus overtly on the nature and organisation of TNCs at the expense of those actors and individuals who are managing the worldwide web of transnational corporations - the entrepreneurs themselves.

There is thus a case for actor-specific studies of Chinese transnational entrepreneurship. This task is particularly daunting in an era of increasing global financial volatility. To a large extent, the success and failure of Chinese business firms abroad are dependent on how entrepreneurial spirits in these firms are constituted and exploited in different host countries and regions. In this paper, I emphasise both **calculated risk-taking** and **foresight** in foreign business ventures. First, although much has been said about the risk-taking behaviour of Chinese entrepreneurs in their 'home' countries in East and Southeast Asia, we still have only limited knowledge of how **transnational** Chinese entrepreneurs exercise their calculated risk-taking behaviour across borders. In my earlier study of Hong Kong firms in Southeast Asia, I found that many Chinese entrepreneurs from Hong Kong tended to activate their transnational social and business networks in order to reduce risks of cross-border operations (Yeung, 1997a; 1997b; 1998a).

As mentioned above, this engagement of transnational networks may **not** be the same phenomenon as the kind of network formation depicted in much of the Chinese business literature. In much of this literature, Chinese entrepreneurs in Southeast Asia were found to rely less on the colonial state to provide the legal enforcement necessary for successful business transactions (Jomo, 1997) than on particularistic ties as a means to achieve 'closure' to outside competitors and to overcome their peculiar form of insecure psyche. Closely-knit networks provided one of the best solutions to overcome these institutional barriers and the personal psyche of fear and insecurity. These networks were based on personal relationships, centered particularly around the family and its immediate circle of social actors (e.g. close friends) (Braadbaart, 1995; Harianto, 1997). As argued earlier, however, many transnational Chinese entrepreneurs today are enrolling non-Chinese actors into their reconfigured transnational business networks in order to penetrate into foreign markets. This is an important change in the risk-taking behaviour of these transnational Chinese entrepreneurs.

Furthermore, good business foresight and acumen are particularly important in ensuring successful transnational operations because of tremendous uncertainty and information asymmetries in the host countries and regions. There is clearly a significant degree of difference between domestic operations and foreign ventures in today's global economy. Transnational Chinese entrepreneurs often need to make major investment decisions under the constraints of imperfect information. Even though some of these entrepreneurs are armed with a stable of professionally-trained analysts and strategists, the final decisions often boil down to a matter of calculated guess. Their capacities in absorbing these enormous risks are highly critical not only in their capabilities to make sound decisions, but more importantly, in their abilities in dealing with crises and failures. Recent empirical events in Asia and the subsequently challenges confronting Chinese business firms are good examples of transnational Chinese entrepreneurship at work.

Globalisation and the Asian Economic Crisis

The above globalisation tendencies aside, the recent Asian economic crisis has potential to impose further serious discipline on the nature and operations of Chinese business firms. Starting with the devaluation of the Thai Baht in August 1997, the financial turmoil has severally affected national economies in Indonesia, Thailand and South Korea. These three different generations of the Asian newly industrialised economies all asked the International Monetary Fund (IMF) for financial bail-out packages. During the second half of 1997 and almost the whole of 1998, many Asian economies have seen their currencies depreciating rapidly against the U.S. Dollar, their stock markets tumbling, their banks and other non-bank financial institutions in serious trouble and their annual growth rates plummeting downwards. The story of the Asian economic crisis is well told elsewhere and will not be repeated here (see Bullard et al., 1998; Haggard and MacIntyre, 1998; Jomo, 1998; Lim, 1998; McLeod and Garnaut, 1998; Radelet and Sachs, 1998; Wade and Veneroso, 1998).

What interest us are the immediate implications of this economic meltdown for Chinese capitalism in Asia. First, it is likely that IMF-driven economic reforms in the region are set to accelerate market liberalisation and deregulation which may result in a more competitive business environment for Chinese firms. Second, the credit squeeze afflicting Asia under austerity measures effectively forces Chinese and non-Chinese Asian firms to increase their transparency to raise funds in international financial markets in order to remain solvent. The lack of transparency in many Asian Chinese business firms is traditionally linked to family control and succession. For many international investors and institutional fund managers, this inward-looking business organisation among ethnic Chinese communities certainly needs to be reconfigured before their credibility and reputation as Asia's leading companies can be restored. This then implies that as an outcome of the economic crisis in Asia, existing competitive pressures emanating from globalisation tendencies are effectively forcing Chinese business firms to open their long-guarded business networks to 'outsiders' from different ethnic and geographical origins. In the next section, I attempt to map out this complex reconfiguration of Chinese business firms and their networks in an era of financial volatility and economic crisis.

Strategies of Crisis Management: Chinese Business Firms From Singapore

In this section, I discuss three major strategies of crisis management by Chinese business firms from Singapore, drawing upon empirical data from an ongoing study of the regionalisation of Singaporean firms. My argument is that successful transnational Chinese entrepreneurs are capable of taking pre-emptive measures or implementing drastic strategies to manage possible economic crises.

Some of these strategies may have been implemented well before the outbreak of the recent Asian economic crisis and therefore may not have been directed towards managing this crisis. With hindsight, however, we can see that their implementation was never to be smooth or without significant obstacles. The point here is that only transnational Chinese entrepreneurs with very strong risk-taking appetites and foresight could have devised and implemented such strategies, which eventually turned out to be the key factor in saving their business empires from collapse in the midst of the Asian economic crisis.

Clearly, there are many such strategies of crisis management in both *ex ante* and *ex post* terms. In an *ex ante* sense, many of the strategies could have been implemented by transnational Chinese entrepreneurs without taking the imminent Asian economic crisis into consideration (see examples below). These strategies subsequently could have saved the Chinese business firms from major financial and operational difficulties. They are good examples of transnational entrepreneurship. In an *ex post* sense, some Chinese business firms have surely chosen to implement certain strategies in order to survive the Asian economic crisis (e.g. cost-cutting measures and market diversification). But the main difference here is that these strategies are specifically an outcome of and driven by the crisis. They are therefore not so 'strategic' after all and are not good examples of transnational entrepreneurship.

I want to focus on three specific strategies of globalisation by some leading Chinese business firms from Singapore:

- Geographical diversification of business interests.
- Financial restructuring through tapping into global capital markets.
- Changing modes of foreign market entry.

In both *ex ante* and *ex post* ways, these globalisation strategies have helped these Chinese business firms to weather the negative impact of the Asian economic crisis. Case studies are presented to analyse the strategies of transnational Chinese entrepreneurs and the nature of their transnational entrepreneurship[3].

Going Global: Transnational Entrepreneurship and Geographical Diversification

In fact, going global has recently become one of the key strategies for Chinese business firms to sustain their growth in an era of accelerated global competition (Yeung and Olds, 2000). Faced with the lack of sizeable domestic markets free from state intervention and monopolistic domination, Chinese business firms are increasingly expanding into regional or even global markets. Kao (1993) has termed this phenomenon 'the worldwide web of Chinese business'. To him, 'Chinese-owned businesses in East Asia, the United States, Canada, and even farther afield are increasingly becoming part of what I call the *Chinese commonwealth*' (Kao, 1993, p. 24 original italics; see also Kotkin, 1992). Today, we witness not only many mega Chinese conglomerates in trading, financial and property markets, but also increasingly world-class manufacturing TNCs owned by ethnic Chinese (Hamlin, 1998). Some exemplary cases of globalising Chinese business firms are Li & Fung and Johnson Electric from Hong Kong (Ellis, 1998; Magretta, 1998), the CP Group from Thailand (Brown, 1998; Pananond and Zeithaml, 1998; Yeung, 1999b) and the Acer Group from Taiwan (Hobday, 1998; Mathews and Cho, 1998).

In the case of Chinese business firms from Singapore, going global did not receive much attention from the Singapore government until very recently when the Asian economic crisis had shown that going regional is simply insufficient to ensure Singapore's future growth. Prior to the official launch of Singapore's regionalisation programme in 1993 (see Yeung, 1998b; 1999c), Chinese business firms from Singapore had been operating extensively in the Asian region. Malaysia was and, still, is the major destination for these ethnic Chinese investments from Singapore (Yeung, 1998c). Since 1993, the regionalisation programme has provided a major impetus for government-linked companies to diversify their operations into the regional market. The geographical focus of this programme then was very much the Asian region. The Asian economic crisis, however, has led to a major reorientation in Singapore's outward investment programme (see Yeung, 2000). In a recent report by The Committee on Singapore's Competitiveness, the state has recognised the importance of exploiting the globalisation of its national firms as a geographical diversification strategy:

> One key lesson from the economic crisis is the need for diversification. We need to maintain a judicious balance between the regional and global dependencies of our economy, and diversify our range of economic activities so as to cushion the impact of a slowdown in any particular region (Ministry of Trade and Industry, 1998, p. 60).

I argue that, to a certain extent, Chinese business firms from Singapore had already been globalising their operations well before this latest call by the state. Just how globalised are these firms from Singapore? Some clues can be found in the data obtained from my recent survey of 204 Singaporean TNCs in which some 54 of them (26.5%) are Chinese family-owned. Table 5.1 shows the historical geographies of these 54 Chinese family-owned TNCs from Singapore. It is clear that the globalisation of these firms occurred well before the 1993 launch of Singapore's regionalisation programme. In fact, their subsidiaries in Hong Kong, Indonesia, Malaysia and Other Regions (e.g. South America and Africa) were mostly established prior to 1985. In terms of their geographical spread, these 54 Chinese family-owned TNCs from Singapore are operating mainly in Asia, in particular China and Malaysia which have respectively attracted some 59% and 74% of them. Very few of them are truly global in their geographical scope of operations. Of the four having operations in Europe, only two have operations in North America and Asia. These two Chinese family-owned TNCs from Singapore are therefore truly global in their operations. In terms of number of subsidiaries, the same geographical pattern emerges, where some 87.5% of all 216 subsidiaries are located in Asia, in particular Malaysia (N = 55) and China (N = 77). On average, each Chinese family-owned TNC from Singapore in our sample owns and controls at least four subsidiaries abroad.

How then does the global reach of these Chinese business firms from Singapore influence their performance in the midst of the Asian economic crisis? How do some of them manage the impact of the crisis by reaping benefits from geographical diversification? In order to explore these questions I now turn to a case study of a truly global TNC from Singapore which is owned and controlled by a Chinese transnational entrepreneur and his family members - the Hong Leong Group.

The founder of the Hong Leong Group is the late Kwek Hong Png who came to Singapore from Fujian, China in 1928. Over a period of half a century, he managed to build up a vast business empire starting with trading, then expanding into property, finance and hotels. The Group's Malaysian branch started in 1963 when the late Kwek Hong Png sent his

Table 5.1   **The Historical Geography of 54 Chinese Business Firms from Singapore** (percentage in parentheses)

Regions/ Countries	Mean Year of Establishment	Number of Operating TNCs	Number of Subsidiaries
Southeast Asia	NA	NA	91 (42.1)
Indonesia	1982	15 (27.8)	16 (7.4)
Malaysia	1983	32 (59.3)	55 (25.5)
Thailand	1988	7 (13.0)	7 (3.2)
Philippines	1994	6 (11.1)	6 (2.8)
Others	1994	7 (13.0)	7 (3.2)
East Asia	NA	NA	98 (45.4)
China	1991	40 (74.1)	77 (35.6)
Hong Kong	1981	14 (25.9)	17 (7.9)
Others	1993	4 (7.4)	4 (1.9)
Europe	1991	4 (7.4)	4 (1.9)
North America	1989	6 (11.1)	6 (2.8)
Other Regions	1985	6 (11.1)	17 (7.9)
Total	NA	54 (100)	216 (100)

Source: Author's survey.

brother Kwek Hong Lye to Malaya (from which Singapore was soon to separate) to extend the family's operations there (East Asia Analytical Unit, 1995, p. 332).

My case study focuses on the Kwek family in Singapore which specialises in property development, finance and hotels. In 1994, Kwek Leng Beng (son of the late Kwek Hong Png) took charge of the Hong Leong Group in Singapore after his father's death. Joining his father's business after finishing his law degree in 1963, Kwek initiated the take over of a loss-making listed company (City Developments), in the late 1960s and early 1970s, and successfully turned it around to become a leading property developer in Singapore. The Hong Leong Group (see Figure 1) is now one of the largest Chinese business groups in Singapore with a market capitalisation value of US$16 billion, an employment strength of 30,000 worldwide and a stable of 300 companies, including 11 listed on various bourses in Singapore, Hong Kong, New Zealand, Manila, New York and London (*The Sunday Times*, 2 February 1997). Through its private property arm (Hong Leong Holdings) and its industrial company (Hong Leong Corporation), Kwek controls China's largest refrigerator manufacturer (Xin Fei Electric) and New York-listed diesel engine maker China Yuchai International (*The Straits Times*, 9 August 1997, p. 10).

Since taking over from his father in late 1994, Kwek has developed a voracious appetite for major acquisitions abroad. The Hong Leong Group has recently globalised into the hotel business through its property development arm - City Developments Ltd. CDL Hotels International, controlled by City Developments Ltd. and listed in the Hong Kong Stock Exchange, manages the group's hotel interest. Since the late 1980s, CDL Hotels International has grown tremendously from owning only 5 hotels in 1989 to 66 in 1998. CDL Hotels International now has a hotel empire spanning 12 countries in Europe, the U.S., Australia, New Zealand, East and Southeast Asia. It is now the eighth largest hotel owner and operator in the world (Annual Report, 1997). Its success in acquiring hotels on a global scale rests not only with the sheer financial muscle of the Hong Leong Group, but more importantly, with the excellent business acumen and transnational entrepreneurship of its present chairman, Kwek Leng Beng.

What then is so entrepreneurial about Kwek Leng Beng to earn him the title of Singapore's Businessman of the Year in 1996? In a rare interview, Kwek said with his characteristic understatement that 'A lot of people say: "You are either born an entrepreneur or you are not. You cannot learn to become one". This is not true' (Cited in *The Sunday Times*, 2 February 1997, p. 2). In fact, Kwek Leng Beng attributed much of his entrepreneurship to the training of his late father Kwek Hong Png:

> It is my strong belief that when you are in it for some time, you can get the feel of it, you get the feel in your guts... I have been trained by my old man and I graduated from his university. He had a lot of strengths, which I learnt over the years. I have been the type of person who likes to soil his hands. I have gone through the process, so I understand the A to Z of the business (cited in *The Sunday Times*, 2 February 1997, p. 2).

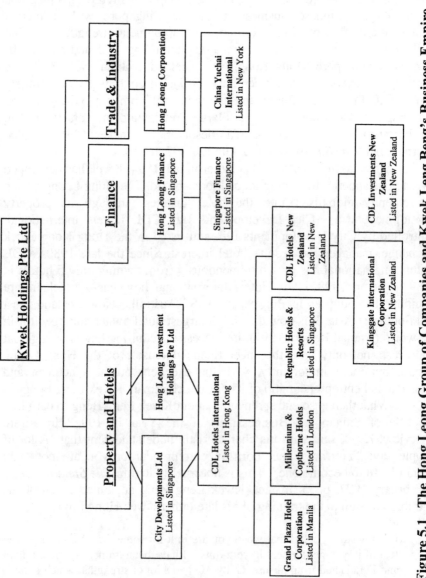

Figure 5.1   The Hong Leong Group of Companies and Kwek Leng Beng's Business Empire

The three important business principles Kwek learnt from his late father were:

- Never overpay for something, particularly when the market has gone crazy.
- Work hard to realise good profits rather than accepting just plain sailing.
- Do something in a big way without over-stretching oneself.

Kwek's superior skills and entrepreneurial instincts in take-over and acquisitions evolved from his early involvement in the take-overs of City Developments Ltd. in the late 1960s and Singapore Finance in 1979. Both listed companies are major contributors to the Hong Leong Group today. In a series of major hotel acquisitions in the 1990s enabling CDL Hotels International to become one of the largest global hotel chains, Kwek has demonstrated tremendous strength and transnational entrepreneurial acumen. He has been credited with an astute eye for choice hotels at bargain prices, often picking them up at rock-bottom prices from receivers. This is not an easy task at all because the transnational entrepreneur needs a lot of experience and good judgement in the absence of complete information about these major assets abroad.

Moreover, Kwek is clearly a risk-taker and he does not like to minimise risks by following the crowd. In another recent interview on why he did not follow the regionalisation strategy of most Singaporean firms, Kwek reflected on his global acquisition trail:

> I think like this: Yes, external wing is good. But when everybody is going to the region, developing properties, that is the time, honestly, I want to sell. I don't follow herd instinct, I fight it... It's common sense. A lot of people thought, it's [Asian markets] a pot of gold, if you go, I must go. But if there are five of you building hotels in the same place, then you won't make it, easy to see... When I went to London to buy hotels, everybody said, 'this guy is talking rot, talking rubbish', but I did not listen. I smelled the market, I know (cited in *The Straits Times*, 20 November 1998, p. 74).

The Kwek way is not to rely on partners, whether they are from Singapore or in the host countries. Very often, Kwek is asked for good projects by interested parties both in and outside Singapore. For example, instead of Kwek being driven by the Singapore government's call for regionalisation since 1993, it is the Singapore government which, through

Temasek Holdings, asked Kwek for good projects, came to evaluate Kwek's suggestions and got involved (e.g. the Beijing Riviera residential project). This is how the Hong Leong Group helps government-linked companies from Singapore to regionalise their operations. According to my interview with a key Kwek family member in Hong Kong (11 June 1998), 'they [government-linked companies] have money, but no expertise in running projects'.

In consequence, Kwek's transnational entrepreneurship and globalisation strategy has paid off in the midst of the Asian economic crisis. All three major listed arms of the Hong Leong Group achieved significant net profits in 1998, a year in which most listed companies in Singapore suffered losses (see Table 5.2). City Developments Ltd., which controls CDL Hotels International, ranks as the 7th most profitable listed company in Singapore in 1998, with a net profit of S$123.7 million. Hong Leong Finance, Singapore Finance and Hong Leong Asia each pocketed a net profit of S$53.2 million, S$32.8 million and S$17 million respectively. The performance of City Developments Ltd. was indeed way ahead of other listed property developers in Singapore. As shown in Table 5.2, among five top losing listed ompanies in 1998, four belonged to property developers (Keppel Land, DBS Land, Orchard Parade Holdings and MCL Land). Four other property developers in Singapore also suffered losses in 1998. One of the major contributing factors to the relative unscathed performance of City Developments Ltd. is the net incomes from its global hotel operations which exceeded those from its core property development business. In 1997, its hotel business in the UK and the US were the most profitable operations and contributed to some 71.8% of profit before taxation by CDL Hotels International Ltd (Annual Report, 1997, p. 75). In 1998, its hotel business turned in profit before interest and tax of S$237.9 million, not far behind the S$269.6 million from both property development and rental income from investment properties.

Kwek Leng Beng said that '[t]his reflects the success of our group's strategy of diversification into international hotels embarked in the early 1990s to ensure a wider spread of earnings' (Cited in *The Straits Times*, 24 March 1999, p. 58). My interview with a key Kwek family member in Hong Kong (11 June 1998) also shows that the Hong Leong Group had been prepared for cyclical fluctuations in the property markets well before the Asian economic crisis. All foreign projects of Hong Leong are run as a financially independent unit and hedging is used to reduce foreign exchange risks. My interviewee also said that Kwek had been telling them

Table 5.2   Performance of Selected Public Listed Companies in Singapore, 1997 and 1998

		Net Profit/(Loss) in S$million		Turnover in S$million	
Ranking	Company	1998	1997	1998	1997
Profitable					
1	OCBC	425.3	581.1	4747.1	4079.1
2	UOB	367.8	502.0	3561.5	3331.2
3	DBS	222.7	436.4	5407.3	3689.0
4	OUB	180.4	254.8	2944.0	2723.6
5	ST Engineering	154.7	120.8	1661.7	1476.7
6	Asia Food & Properties	140.9	45.2	954.9	1432.8
7	**City Developments**	**123.7**	**409.2**	**2043.3**	**2470.4**
15	**Hong Leong Finance**	**53.2**	**72.9**	**422.8**	**374.9**
22	**Singapore Finance**	**32.8**	**34.8**	**153.1**	**135.9**
37	**Hong Leong Asia**	**17.0**	**(34.8)**	**456.3**	**616.0**
Loss-Making					
1	NOL	(438.2)	(297.3)	6485.3	2672.5
2	Keppel Land	(350.6)	104.6	317.9	621.2
3	DBS Land	(239.0)	182.3	1419.8	1083.2
4	Orchard Parade Holdings	(196.2)	21.6	193.9	491.2
5	MCL Land	(188.5)	19.5	297.0	297.0
9	Tuan Sing	(83.9)	31.6	575.9	646.3
43	United Overseas Land	(4.1)	52.6	425.5	357.6
-	Wing Tai[1]	(99.8)	46.0	-	-
-	First Capital[1]	(17.8)	6.3	-	-

[1] Data refer to 6 months ended December 1997 and 1998.
Source: *The Straits Times*, 24 March, 1999, p. 58; 2 April, 1999, p. 86.

that 'the market is not going up every day. We must also be prepared for the worst'.

Hong Leong's London-listed subsidiary, Millennium & Copthorne Hotels (M&C), was CDL Hotels International's star performer in 1998. M&C was listed in the London Stock Exchange on 25 April 1996, slightly more than one year before the outbreak of the Asian economic crisis. The listing attracted some £1.2 billion investor funds worldwide. Kwek's commented a day before M&C's listing that '[w]e are very happy with the timing. It could not have been more perfect from the beginning up to the flotation' (Cited in *The Straits Times*, 24 April 1996). Indeed, this comment could not have been more appropriate in the context of the crisis. The listing of M&C not only generated net proceeds of £174.5 million to relieve both M&C and its parent company CDL Hotels International from debt obligations, but also significantly raised the investment profile of Kwek in major global capital markets. This latter point proved to be important when Kwek needed to tap into these capital markets in the midst of the Asian economic crisis. In brief, what was initially an egg-spreading strategy pursued relentlessly by Kwek has turned out to be the key to shield the Group from the severe impact of the Asian economic crisis.

Financial Restructuring: Tapping Into Global Capital Markets

Tapping into the global capital markets is another important strategy to manage the impact of the Asian economic crisis. It is also a pre-emptive measure in that those Chinese business firms which had implemented such a strategy tended to be less badly affected by the crisis. To date, many Chinese business firms continue to rely on their social networks and personal ties in accessing funds and capital for investment. Indeed, many of today's renowned Chinese entrepreneurs in Asia started their businesses by pooling capital from various network sources (e.g. Redding, 1990; Chan and Chiang, 1994; Backman, 1999).

However, as their business empires have grown and their scale of operations has extended across national boundaries, many of these transnational Chinese entrepreneurs no longer restrict their sources of capital exclusively to social and business networks embedded in their 'home' countries. The globalisation of their business activities is also fuelled by their ability to tap into global capital markets. Though not many of these Chinese business firms have been successful in tapping into global capital funds (see Olds and Yeung, 1999), those who managed to do so have tended to weather the recent Asian economic crisis better. This being the case, the crisis presented golden opportunities for these transnational Chinese entrepreneurs to consolidate and expand their business empires within and outside the Asian region. What then are the critical components of their strategies in tapping into global capital markets?

First, these transnational Chinese entrepreneurs must be able to transfer the goodwill extensively developed in their 'home' countries. Their goodwill often emanates from their strong embeddedness in domestic financial and business networks. Many of them are also involved in the banking and finance industry (e.g. Liem Sioe Liong in Indonesia's Bank Central Asia; Kwek Leng Beng in Singapore's Hong Leong Finance and Wee Cho Yaw in Singapore's United Overseas Bank; Quek Leng Chan in Malaysia's Hong Leong Finance and Hong Kong's Dao Heng Bank). In some Southeast Asian countries, this goodwill also originates from their extensive involvement in political-economic alliances with leading politicians and military personnel (McVey, 1992; Brook and Luong, 1997; Hefner, 1998).

This strong embeddedness in goodwill of a domestic origin presents a significant problem for the globalisation of these Chinese business firms. To what extent can this goodwill and its related competitive advantage be transferred to other geographical contexts (see also Hu, 1995)? Going back to the case study of Kwek Leng Beng's Hong Leong Group, it is clear that Kwek has enjoyed tremendous goodwill transfer in his global operations. He has successfully listed many of his hotel businesses in stock exchanges in Hong Kong, London and New Zealand. As mentioned above, his successful placement of Millennium & Copthorne Hotels in the London Stock Exchange has not only brought him immediate financial gains, but also energized the London capital market for hotel investments (*The Straits Times*, 24 April 1996). In 1997, Prudential Client (MSS) Nominees Ltd. was a substantial minority shareholder of M&C Hotels Plc. (Annual Report, 1997, p. 23). Kwek's CDL Hotels International, listed in the Hong Kong Stock Exchange, has also received very good support from major banks in Hong Kong (Interview with a key Kwek family member in Hong Kong, 11 June 1998).

Second, those Chinese business firms which have hedged their long-term borrowings and diversified their bankers have tended to manage better the major impact of the Asian economic crisis. In the midst of the crisis, foreign exchange risks and liquidity squeeze were the major problems confronting many Asia firms. For the Korean *chaebols*, the problem rests with their high debt-equity ratios (Mathews, 1998; Wade and Veneroso, 1998). For many Indonesian firms, liquidity squeeze became a major problem when many Indonesian banks were put under the disciplinary programme imposed upon Indonesia by the IMF (Bullard et al., 1998; Backman, 1999). Similarly, rapidly falling stock market prices meant that these firms were worth much less than their collateral would have indicated, if there ever were such collateral!

The important lesson here is that had these firms, particularly those with international business exposure, hedged and diversified their long-term borrowings, they would have been much better at riding out the crisis. In the case of CDL Hotels International, hedging of funds is always used as an important instrument to minimize foreign exchange exposure (Interview with a key Kwek family member in Hong Kong, 11 June 1998). The breakdown of CDL Hotels International's long-term borrowings, as at 31 December 1997, indicates that its HK$5.9 billion borrowings were denominated in the domestic currencies of the U.S., the U.K., New Zealand, Malaysia, Singapore and Taiwan (Annual Report, 1997). Some 60% of these borrowings were denominated in Sterling pounds alone, indicating the importance of London-based financial institutions in supporting the globalisation of CDL Hotels International.

In addition to hedging long-term borrowings, Kwek has enjoyed a highly diversified portfolio of bankers for both his CDL Hotels International and M&C Hotels Plc. These diverse range of bankers for both listed companies in Hong Kong and London not only illustrates their strong financial support for Kwek's ambition to globalise his hotel business, but also reduces his risk of liquidity squeeze in times of economic crisis. For example, CDL Hotels International has amongst its bankers those from the U.K., France, Japan, Hong Kong, Switzerland, Australia and New Zealand. Its subsidiary, M&C Hotels Plc, has also gained strong support from four major British banks - National Westminster Bank, Midland Bank, ING Barings and HSBC and Lloyds Bank. With diversified sources of capital, it is not surprising that both CDL Hotels International and M&C Hotels Plc have been able to weather major financial constraints in the midst of the Asian economic crisis. The case demonstrates the importance of globalising into different capital markets as a key proactive strategy for the long term expansion of Chinese business firms across borders. The recent economic crisis has just made the point even clearer and more relevant.

Changing Modes of Market Entry: From Direct Investments to Non-Equity Investments

The globalisation of Chinese business firms takes a great variety of organisational forms. The choice of different modes of entry into foreign markets becomes an important issue not only in understanding their internationalisation processes (see Yeung, 1999b), but also in assessing their capabilities in meeting the challenges of the recent Asian economic crisis. In general, there are many ways of organising transnational operations, from arm's-length market transactions (e.g. exports) to fully-integrated vertical hierarchies (e.g. FDI). Before the Asian economic crisis, joint ventures and acquisitions were perhaps the most common organisational modes through which Chinese business firms internationalised their operations. In fact, many Chinese business firms on entering China joined hands with local enterprises and state institutions (e.g. the CP Group from Thailand). The more opaque business environments in China and most Southeast Asian countries favoured joint ventures between Chinese business firms and indigenous enterprises/state institutions. These Chinese business firms might also team up with Western firms in order to establish themselves in some Asian countries. For example, Chinese business firms from Singapore have long been promoting themselves as the 'gateway' to the East for the West.

Acquisitions were also preferred by large Chinese business conglomerates to control their foreign subsidiaries (e.g. the Salim Group from Indonesia). When these firms globalised into regions outside Asia, acquisitions became an even more important instrument to enter into these unfamiliar foreign markets. Rather than greenfield operations, acquisitions were preferred because of speed and risk factors. The open business environment in North America and Western Europe does not give these Chinese business firms any competitive advantage which they previously enjoyed in the more opaque Asian business environments. Rather, these open business fields in the West (e.g. high-tech industries and hotel businesses) tend to promote intensified competition which accentuates the importance of economies of scale, technological innovations and strong expertise, particularly to those new entrants from Asia. Acquisitions of existing operations in these countries facilitate risk minimisation, experience building and major subsequent investments in the host countries. The best example from Singapore is Kwek Leng Beng's in-roads into global hotel businesses through a series of major acquisitions in the U.K., the U.S. and New Zealand.

In the context of the Asian economic crisis, direct investments through either greenfield operations or acquisitions may not be suitable modes of entry into the highly risky regional markets because of financial constraints. Banks and financial institutions are more reluctant to finance major foreign acquisitions by Chinese business firms from Asia. Instead, they prefer these Chinese business firms to consolidate their positions and, sometimes, to repay their debts before incurring more new debts. Chinese business firms from Singapore therefore must look into other modes of non-equity investments abroad, including franchising, management contracts, cooperative agreements, licensing, subcontracting, strategic alliances and so on. These non-equity modes of foreign market entry can reduce risk and capital commitment typical in other forms of direct investments. There is also sector-specificity in the use of these different non-equity modes of investments. Whereas manufacturing firms may prefer subcontracting, licensing and strategic alliances, service firms may find franchising and management contracts particularly attractive in both minimising capital outlays and securing market share in the host countries.

SINFood, a leading Chinese dried-food manufacturing and retail family business, is using franchising precisely as the key means for regionalising into Asia and eventually globalising into North America. Founded in 1945, SINFood is now owned and managed by the third generation of the Wong family. It had a relatively significant annual turnover of US$20.6 million, fixed assets of US$14.7 million and a global workforce of more than 500 in 1997 (Interview with Mr. Wong in Singapore, 6 May 1998). Its long-term objective is 'to become not only the recognised leader in the [dried food] trade in Asia with an international chain of outlets, but also in other traditional Chinese delicacies' (Company Profile, 1995, p. 10).

Today, SINFood has a 35% global market share in Chinese foodstuffs. This successful market penetration is partly explained by the relentless pursuit of geographical diversification and franchising by the company under its current third-generation leadership. Mr. Wong, Managing Director of SINFood, is credited with introducing the concept of franchising as the key means of growing the company beyond the domestic market into Hong Kong, China and Malaysia. Before he took over the management, his uncles had focused very much on retail business in the Singapore market. As he explained, '[w]e're a family business. So we didn't have a formal management. Our main business really took off from the third generation' (Interviewed in Singapore, 6 May 1998). As envisioned in his company profile in 1995,

With our experience and proven business system, we have begun realising this vision, by transferring our business know-how to other parts of the world through franchising. Our franchise marketing programme, initially targeted at Hong Kong and China, will eventually reach as far as Canada and the United States of America... Diversification and franchising are just the first steps in [SINFood's] corporate evolution, and we will continue to evolve our operations and products to suit the changing needs and tastes of our customers. In line with our globalisation strategy, we are also working toward listing [SINFood] on the stock exchange within this decade.

Our key concern here is how SINFood manages the impact of the Asian economic crisis through transnational entrepreneurship and continuous engagement in franchising abroad. To overcome the difficulties in the midst of the crisis, a transnational entrepreneur needs more than vision and foresight; he/she also needs perseverance and commitment to the foreign ventures. Mr. Wong had always wanted to go beyond his inherited business and to establish his transnational empire of Chinese dried food outlets. Although the crisis put pressures on his operations abroad, he was determined to maintain his commitment to a regional franchising business. SINFood's franchising programme has therefore not only enabled it to establish a strong foothold in the regional market, but also helped it to escape narrowly a heavy capital write-off, as a consequence of direct capital investments in retail outlets abroad. As Mr. Wong noted, 'This economic crisis changes predictions. For example, we predicted quite low figures for these 2 years because of low confidence. It's good enough if we could maintain the figures!'. Wong's entrepreneurial foresight, i.e. not over-stretching his financial resources abroad through aggressive direct investments, is clearly an important lesson for such Chinese family firms as SINFood on how to survive the Asian economic crisis.

Conclusion and Implications for Research and Policies

This paper has emphasised the role of transnational Chinese entrepreneurship in managing crisis tendencies in an era of accelerated globalisation. Whether they choose to participate in globalisation as local competitors to global corporations and/or as global competitors in their own right, these emerging ethnic Chinese-owned and controlled TNCs are subject to the tyranny and game rules of global competition. Though still heavily biased towards 'quick money' sectors, their business fields are growing wider to incorporate both high tech industries and producer services; though still largely dominated by family members, their management structures are highly professionalised; though still constituted by many ethnic Chinese capitalists, their business networks are being reshaped in a globalising era by non-Chinese actors who are capable of connecting these transnational Chinese entrepreneurs to global capital markets in London and New York. Based on a survey of Chinese family business in Singapore and qualitative case studies, I have shown three major pre-emptive strategies through which these firms have managed the shocks of the recent Asian economic crisis. Though not exclusive to the range of potential strategies for crisis management, I consider as most proactive these three strategies of geographical diversification, tapping into global capital markets and using non-equity investments to penetrate foreign markets.

My discussion on these proactive strategies clearly opens a whole range of issues for debate which may have important implications for research and policies on the globalisation of Chinese business firms (see also Yeung and Olds, 2000). I want to address three key implications here:

- Access to capital.
- Internal management structures and processes.
- Sources of dynamic competitive advantage.

First, as amplified by the Asian economic crisis, sources of capital play a significant role in influencing how Chinese business firms can meet the challenges of globalisation, both as local competitors and/or as globalising firms. For researchers, it is important to go beyond a study of how Chinese business firms pool together capital based on locally constituted social and business networks (e.g. bank financing). We need to examine how large-scale Chinese business firms are tapping into capital markets outside their 'home' countries. We need to know more about how they transfer their goodwill to other localities and how they embed themselves in business networks abroad. Empirical studies of how Li Ka-shing managed to transfer his goodwill to tap into local business networks in Vancouver are clearly a good starting point (see Mitchell, 1995; Olds, 1998).

Second, globalising trust and credibility requires Chinese business firms to pay a lot more attention to their internal management structures and processes than would be the case in domestically-oriented Chinese business firms. During the 1997/1998 Asian economic crisis, many Asian firms failed because they were so poorly managed that even a cheap sale of their assets did not attract foreign investors. Many of these ailing Asian firms were also family firms, though the phenomenon might not be exclusive to Chinese family business. Embracing globalisation implies more than buying and selling assets in other countries beyond one's home turf. It is also about how they can manage these foreign assets and/or advantages better than their competitors. Being a family business does not necessarily mean that it cannot be professionalised. Indeed, there are no inherent limits to the growth and professionalisation of Chinese business firms (Yeung, 1999d). The challenge to researchers is to identify the best management practices which can contribute to the successful globalisation of these once highly patriarchal firms. We must also situate these best practices in their social and institutional contexts.

Third, the point of sustainability is important in understanding the dynamics of the competitive advantage of Chinese business firms. With globalisation, Chinese business firms must actively search for new sources of dynamic competitive advantage. Many of these firms have grown from imperfect and relatively monopolistic domestic markets. They have often enjoyed tremendous advantages in these domestic markets because of their special connections and relationships with ruling politicians and/or key business elites. Special licenses and monopoly rights have been granted to these Chinese business firms which become their major 'cash cows'. When these firms venture into foreign markets, the scenario is almost completely different. The field of competition becomes much more open and level; only the fittest and most competitive firms will survive. Globalising Chinese business firms is clearly not an easy task, but it is surely one of the best proactive ways to manage crisis tendencies in a era of accelerated globalisation.

Notes

[1] The term 'overseas Chinese' may be contentious to some scholars of ethnic Chinese living outside mainland China. See Wang (1991) and Lynn (1998) for two authoritative accounts of the origin and status of ethnic Chinese living outside mainland China. Throughout this paper, I generally refer to 'ethnic Chinese' or to specific groups (e.g. Hong Kong entrepreneurs) rather than 'overseas Chinese' in the discussion of research materials. But references to the literature sometimes require reference to 'overseas Chinese' to be clear.

[2] I use 'home' here because many Southeast Asian countries might not be the birth place for the first and, sometimes, second generations of many of these transnational Chinese entrepreneurs.

[3] The choice of case studies in this paper has been made made for two reasons. First, they must have been relatively successful in weathering the Asian economic crisis, i.e. satisfying construct validity. Second, they are selected on the basis of the completeness of their available information. All case studies here are discussed for illustration purposes. I have no intention nor belief that these case studies can provide universal generalisation and invariant laws (see Yin, 1994; Numagami, 1998).

References

Backman, M. (1999), *Asian Eclipse: Exposing the Dark Side of Business in Asia*, John Wiley, Singapore.

Braadbaart, O. (1995), 'Sources of ethnic advantages: a comparison of Chinese and *pribumi*-managed engineering firms in Indonesia', in R.A. Brown (ed.), *Chinese Business Enterprise in Asia*, Routledge, London, pp. 177-96.

Brook, T. and Luong, H.V. (eds.) (1997), *Culture and Economy: The Shaping of Capitalism in Eastern Asia*, University of Michigan Press, Ann Arbor.

Brown, R.A. (ed.) (1995), *Chinese Business Enterprise in Asia*, Routledge, London.

Brown, R.A. (1998), 'Overseas Chinese investments in China - patterns of growth, diversification and finance: the case of Charoen Pokphand', *The China Quarterly*, no. 155, pp. 610-36.

Bullard, N., Bello, W. and Mallhotra, K. (1998), 'Taming the tigers: the IMF and the Asian crisis', *Third World Quarterly*, vol. 19, no. 3, pp. 505-55.

Carney, M. (1998), 'A management capacity constraint? Obstacles to the development of the overseas Chinese family business', *Asia Pacific Journal of Management*, vol. 15, pp. 137-62.

Chan, K.B. and Chiang, S.N.C. (1994), *Stepping Out: The Making of Chinese Entrepreneurs*, Simon and Schuster, Singapore.

Dicken, P. (1998), *Global Shift: Transforming the World Economy*, Paul Chapman, London.

Dunning, J.H. (1988), *Explaining International Production*, Unwin Hyman, London.

Dunning, J.H. (1993), *Multinational Enterprises and the Global Economy*, Addison Wesley, Reading, MA.

East Asia Analytical Unit (1995), *Overseas Chinese Business Networks in Asia*, Department of Foreign Affairs and Trade, Parkes, Australia.

Ellis, P. (1998), 'Johnson Electric', *Asian Case Research Journal*, vol. 2, pp. 53-66.

Fukuyama, F. (1995), *Trust: The Social Virtues and the Creation of Prosperity*, Hamish Hamilton, London.

Haggard, S. and MacIntyre, A. (1998), 'The political economy of the Asian economic crisis', *Review of International Political Economy*, vol. 5, no. 3, pp. 381-92.

Haley, G.T., Tan, C.T. and Haley, U.C.V. (1998), *The New Asian Emperors: The Overseas Chinese, Their Strategies and Competitive Advantages*, Butterworth-Heinemann, Oxford.

Hamilton, G.G. (ed.) (1991), *Business Networks and Economic Development in East and South East Asia*, Centre of Asian Studies, University of Hong Kong, Hong Kong.

Hamlin, M.A. (1998), *Asia's Best: The Myth and Reality of Asia's Most Successful Companies*, Prentice Hall, Singapore.

Harianto, F. (1997), 'Business linkages and Chinese entrepreneurs in Southeast Asia', in T. Brook and H.V. Luong (eds.), *Culture and Economy: The Shaping of Capitalism in Eastern Asia*, University of Michigan Press, Ann Arbor, pp. 137-53.

Hefner, R.W. (ed.) (1998), *Market Cultures: Society and Values in the New Asian Capitalisms*, Institute of Southeast Asian Studies, Singapore.

Hobday, M. (1998), 'Latecomer catch-up strategies in electronics: Samsung of Korea and Acer of Taiwan', *Asia Pacific Business Review*, vol. 4, nos. 2/3, pp. 48-83.

Hodder, R. (1996), *Merchant Princes of the East: Cultural Delusions, Economic Success and the Overseas Chinese in Southeast Asia*, John Wiley, Chichester.

Hsing, Y.T. (1998), *Making Capitalism in China: The Taiwan Connection*, Oxford University Press, New York.

Hu, Y.S. (1995), 'The international transferability of the firm's advantages', *California Management Review*, vol. 37, no. 4, pp. 73-88.

Jomo, K.S. (1997), 'A specific idiom of Chinese capitalism in Southeast Asia: Sino-Malaysian capital accumulation in the face of state hostility', in D. Chirot and A. Reid (eds.), *Essential Outsiders: Chinese and Kews in the Modern Transformation of Southeast Asia and Central Europe*, University of Washington Press, Seattle, pp. 237-57.

Jomo, K.S. (ed.) (1998), *Tigers in Trouble: Financial Governance, Liberalisation and Crises in East Asia*, Zed Books, London.

Kao, J. (1993), 'The worldwide web of Chinese business', *Harvard Business Review*, March-April, pp. 24-36.

Kotkin, J. (1992), *Tribes: How Race, Religion and Identity Determine Success in the New Global Economy*, Random House, New York.

Lever-Tracy, C., Ip, D. and Tracy, N. (1996), *The Chinese Diaspora and Mainland China: An Emerging Economic Synergy*, Macmillan, London.

Lim, L.Y.C. (1998), 'Whose 'model' failed? Implications of the Asian economic crisis', *Washington Quarterly*, vol. 21, no. 3, pp. 25-36.

Lim, L.Y.C. and Gosling, L.A.P. (eds.) (1983), *The Chinese in Southeast Asia*, Maruzen Asia, Singapore.

Lim, L.Y.C. and Gosling, L.A.P. (1997), 'Strengths and weaknesses of minority status for Southeast Asian Chinese at a time of economic growth and liberalization', in D. Chirot and A. Reid (eds.), *Essential Outsiders: Chinese and Kews in the Modern Transformation of Southeast Asia and Central Europe*, University of Washington Press, Seattle, pp. 285-317.

Lynn, P. (ed.) (1998), *The Encyclopedia of Chinese Overseas*, Archipelago Press, Singapore.

McLeod, R.H. and Garnaut, R. (eds.) (1998), *East Asia in Crisis: From Being a Miracle to Needing One?*, Routledge, London.

McVey, R. (ed.) (1992), *Southeast Asian Capitalists*, Cornell University Southeast Asia Program, Ithaca.

Magretta, J. (1998), 'Fast, global, and entrepreneurial: supply chain management, Hong Kong style: an interview with Victor Fung', *Harvard Business Review*, vol. 76, no. 5, pp. 103-14.

Mathews, J.A. (1998), 'Fashioning a new Korean model out of the crisis: the rebuilding of institutional capabilities', *Cambridge Journal of Economics*, vol. 22, no. 6, pp. 747-59.

Mathews, J.A. and Cho, D.S. (1998), *Tiger Chips: The Creation of a Semiconductor Industry in East Asia, 1975-2000*, Cambridge University Press, Cambridge.

Ministry of Trade and Industry (1998), *Committee on Singapore's Competitiveness*, MTI, Singapore.

Mitchell, K. (1995), 'Flexible circulation in the Pacific Rim: capitalism in cultural context', *Economic Geography*, vol. 71, no. 4, pp. 364-82.

Numagami, T. (1998), 'The infeasibility of invariant laws in management studies: a reflective dialogue in defense of case studies', *Organization Science*, vol. 9, no. 1, pp. 2-15.

Olds, K. (1998), 'Globalization and urban change: tales from Vancouver via Hong Kong', *Urban Geography*, vol. 19, no. 4, pp. 360-85.

Olds, K. and Yeung, H.W.C. (1999), '(Re)shaping 'Chinese' business networks in a globalising era', *Environment and Planning D: Society and Space*, vol. 17, no. 5, pp. 535-55.

Pananond, P. and Zeithaml, C.P. (1998), 'The international expansion process of MNEs from developing countries: a case study of Thailand's CP Group', *Asia Pacific Journal of Management*, vol. 15, pp. 163-84.

Radelet, S. and Sachs, J.D. (1998), 'The East Asian financial crisis: diagnosis, remedies, prospects', *Brookings Papers on Economic Activity*, vol. 1, pp. 1-90.

Redding, S.G. (1990), *The Spirit of Chinese Capitalism*, De Gruyter, Berlin.

Smart, A. and Smart, J. (1998), 'Transnational social networks and negotiated identities in interactions between Hong Kong and China', in M.P. Smith and L.E. Guarnizo (eds.), *Transnationalism From Below*, Transaction Publishers, New Brunswick, pp. 103-29.

The Straits Times. Singapore, various issues.

The Sunday Times. Singapore, 2 February 1997.

Wade, R. and Veneroso, F. (1998), 'The Asian crisis: the high debt model versus the Wall Street-Treasury-IMF complex', *New Left Review*, no. 228, pp. 3-23.

Wang, G. (1991), *China and the Chinese Overseas*, Times Academic Press, Singapore.

Weidenbaum, M. and Hughes, S. (1996), *The Bambook Network: How Expatriate Chinese Entrepreneurs Are Creating a New Economic Superpower in Asia*, The Free Press, New York.

Yeung, H.W.C. (1997a), 'Business networks and transnational corporations: a study of Hong Kong firms in the ASEAN region', *Economic Geography*, vol. 73, no. 1, pp. 1-25.

Yeung, H.W.C. (1997b), 'Cooperative strategies and Chinese business networks: a study of Hong Kong transnational corporations in the ASEAN region', in P. W. Beamish and J. P. Killing (eds.), *Cooperative Strategies: Asia-Pacific Perspectives*, The New Lexington Press, San Francisco, CA, pp. 22-56.

Yeung, H.W.C. (1998a), *Transnational Corporations and Business Networks: Hong Kong Firms in the ASEAN Region*, Routledge, London.

Yeung, H.W.C. (1998b), 'The political economy of transnational corporations: a study of the regionalisation of Singaporean firms', *Political Geography*, vol. 17, no. 4, pp. 389-416.

Yeung, H.W.C. (1998c), 'Transnational economic synergy and business networks: the case of two-way investment between Malaysia and Singapore', *Regional Studies*, vol. 32, no. 8, pp. 687-706.

Yeung, H.W.C. (1999a), 'Under siege? Economic globalisation and Chinese business in Southeast Asia', *Economy and Society*, vol. 28, no. 1, pp. 1-29.

Yeung, H.W.C. (1999b), 'The internationalization of ethnic Chinese business firms from Southeast Asia: strategies, processes and competitive advantage', *International Journal of Urban and Regional Research*, vol. 23, no. 1, pp. 103-27.

Yeung, H.W.C. (1999c), 'Regulating investment abroad? The political economy of the regionalisation of Singaporean firms', *Antipode*, vol. 31, no. 3, pp. 245-73.

Yeung, H.W.C. (1999d), 'Limits to the growth of family-owned business? The case of Chinese transnational corporations from Hong Kong', *Family Business Review*, vol. 12.

Yeung, H.W.C. (2000), 'State intervention and neoliberalism in the globalising world economy: lessons from Singapore's regionalisation programme', *The Pacific Review*, vol. 13, no. 1, pp. 133-62.

Yeung, H.W.C. and Olds, K. (eds.) (2000), *The Globalisation of Chinese Business Firms*, Macmillan, London.

Yin, R.K. (1994), *Case Study Research: Design and Methods*, Sage, Thousand Oaks, CA.

6 Networks and Strategies in Taiwanese Business

DAVID IP

Introduction

About ten years ago, central Taiwan was one of the major growth poles on the island. Production in Taichung, Zhanghua, and Nantou, specialising in textiles, shoes, garments, umbrellas and machinery, accounted for more than one-third of Taiwan's total export value. The region had been regarded as one of the significant bases propelling Taiwan's economic miracle. Appropriately, entrepreneurs from the region prided themselves on their innovation, diligence and persistence. Yet beginning in August 1998, the first sign of crisis emerged when the Reilian Consortium (Rei-lian Ji-tuan) experienced financial difficulties and its share prices started to drop rapidly. In early November, the well-established industrial equipment manufacturer Taichung Jing-Ji (TJJ), after spending NT$20 million acquiring 51% of the shares of the Japanese industrial company Lintec, its subsidiary, suddenly found itself in financial trouble and had to apply for government assistance. In the following two months, the share prices of 16 major enterprises in Taichung, Zhanghua and Nantou started to tumble one after another. Some of these enterprises ultimately were sold to new owners and worse still, the CEO (Tsai Kunming) of the only publicly floated company in Nantou (Huan-long Electronics, HLE) was arrested for embezzlement of NT$360 million (about one-fifth of the company's asset value). Concurrently, two of the three major banks from Taichung suffered massive losses (Panasia Bank lost NT$640 million and Taichung Commercial Bank, TCB, was reportedly losing approximately similar amounts). At the end of November 1998, this triggered a run on the Taichung Commercial Bank. In two days, a total of NT$550 billion was withdrawn from the bank and its CEO was consequently sacked after being on the job for only a month (Chan 1998, Chan 1999, Shen 1999).

For quite a while, some Western observers had held the view that the essential characteristics that many diaspora Chinese businesses

114

inherited – flexibility, personalised networks, trust and diversification – and to which others credited Asia's 'miracle economy', were hyped and somewhat dubious. They now have argued that events and stories like the ones we have just outlined, could not have been more convincing proof that they had been right all along – that an economy or business that has been riding its success on nepotism and interference with free market rationality and the transparency of level playing fields will eventually be doomed to fail. In their view, the panacea for healing the ills of these businesses and of the economies in which they operate, will involve swallowing a bitter pill – the complete acceptance of a supposed Western model based on professional management and its values of accountability, credibility and transparency, and a total rejection of network capitalism that promotes secrecy and suspicion (Summers 1998).

Interestingly, the concern over whether personalised networks are the key to their success or failure for diaspora Chinese enterprises has also been similarly mirrored, although not exactly in the same terms, in the debates between the 'social embeddedness thesis' and the 'organisational imperative thesis'. The former, notably originating with Granovetter's (1985) critique of rational and utilitarian analyses of economic behaviour, suggests that any economic transaction tends to involve pre-existing social networks. In this context, networks have always been a central feature of modern economy because of the fundamental problem of trust. Not even contract or any economic device could completely remove the pervasive problems of opportunism and malfeasance. The 'social embeddedness thesis' thus argues that even in the case of business organisations, inter-organisational cooperation is necessary and because of this, they will be embedded in the web of social relations amongst individuals. In turn, this brings into economic transactions influences of social relations, networks of multiple relations of past and present.

The 'organisational imperative thesis', however, argues that in the case of an organisation, although it is necessary to use social ties to secure resources, benefits and cooperation from others, many of these ties are cultivated rather than pre-existent. They are therefore instrumental and incidental rather than fundamental. Particularly significant is that an organisation is designed to have its own interests that may not necessarily coincide with the personal interests of its individual members. When organisational interests are at stake, the organisation, and not the individuals, will reign. The advantages proposed by the 'social embeddedness thesis' are thus seen to be severely limited and less important (Tam 1999).

It seems that both the theories about the strength of diapora Chinese capitalism and the importance of social embeddedness of personalised networks have been put to test by the Asian crisis. It is almost two and a half years down the track since the Asian crisis first struck Thailand. In the period leading up to the crisis how far did social embeddedness facilitate debt, encourage overly rapid growth, inflation of self and business egos and thus lead to bubbles? Is the crisis a result of an over-extension or over reliance on such networking? Has speeding up of network creation been dangerous? How do networks remain operative in crisis? How far have they been crucial to survival, success and new startups since the crisis? In this paper we do not aspire to resolve the arguments in the recent debates as they are substantial and unfinished, and will require much greater efforts than is allowed here in a chapter. Rather we prefer to initiate a new approach to these questions and concerns through a careful review of some case materials of recent failures and successes among Taiwanese businesses. From these we hope to form some preliminary impressions that will help to clarify the relevance, if not the distinctive strength and weaknesses, of the role of social networks in business.

From Social Networks to Production Networks

We have argued elsewhere (Lever-Tracy et al 1996) there have been considerable advantages to diaspora Chinese entrepreneurs in their use of long-term, trust-based, personal relationships and networks linking the members of different owning families when operating their businesses. While pure market rationality may be compromised in the process, the increase in reliable information and the reduction of risk provided by networks, and thus the minimisation of business costs can more than compensate. With them small actors, in particular, are more likely to be more confident about their undertakings. Without them they are also less likely to venture into the unknown, even where the incentive of new opportunities was strong.

A review of the success stories of the Taiwanese businesses is testimony to the demonstrated positive benefits of utilisation of social networks. Xie (1999), for example, in his definitive sweeping analysis of the rise of the southerners or the 'Southern Gang'[1] (Tainan Bang) since 1920s in Taiwan's corporate world, has shown that the present major large corporations that dominate Taiwan's food processing, textiles, cement, finance and retail sectors all had a humble beginning, using only their existing networks to establish, expand and multiply their ventures. Even more striking is his demonstration of how these southern Taiwanese business tycoons are now sitting members of boards directors in each other's enterprises, constituting the particular pattern of ownership as well as business network which Numazaki (1993) specifically termed 'banana-bunch shaped business group'.

Less complicated but decidedly reliant on kinship solidarity is the case of Tainan Enterpises (TE), a garment manufacturing firm that almost faced closure during the late 1980s. It has now rebounded so successfully that when it was listed on the stock market in Taiwan in April 1999, its shares were in such high demand that their prices exceeded those of Acer and Iventec, which were heralded by the media as the 'hot chips'. In 1998, despite the Asian crisis, it scored an annual growth rate of 26.4% and had a workforce of 6,000 in factories spreading from China to Indonesia and Cambodia, producing garments for famous American brand names such as GAP, Banana Republics and Ann Taylor. The owner, Yang Qingfeng attributed his success to family cohesion. He recounted that when he first inherited the business from his father, not only was he confronted with serious problems of labour shortage and of the appreciation of the Taiwanese currency against the US Dollar, worse still, his manufacturing plants were destroyed by fire first in 1984, and again the following year. What gave him the strength to rebuild his business were his siblings. He took charge of the business looking after its overall development, his brother-in-law oversaw the operation, and his two sisters took over marketing and finance. In his view, 'the key to running a stable and successful family business is to delegate your family members so that they could exercise their best capability, but you also provide them with a sense of security [because of being in the same family]' (Guan 1999, p. 66).

Likewise, Lin Bai-li, in a recent interview, acknowledged the 'inputs' he received from the old boy network he had established when he was a student at the age of 23, in 1972. Lin is owner of Quanta Computer Company Limited, a manufacturer of notebook computers for Dell and Apple, which also had the most spectacular rise, from nowhere in 1998 to number five in 1999, on the annual listing of the largest 500 Chinese business corporations in Asia (*Yazhou Zhoukan*)[2]. One of his classmates was Wen Xiren, now deputy director of Inventec Corporation, ranked 10th as Asia's largest computer manufacturer, specialising also in notebook computers. Wen and Lin later became partners in their own computer company called San-Ai Electronics. The present director of Inventec, Ye Guoyi, was one of the team members employed in San-Ai. Fifteen years later when Lin decided to go it alone in establishing Quanta, both Wen and Ye contributed one-fourth of the venture capital. 'When Quanta was first started, the total capital involved was NT$ 80 million. They [Wen and Ye] contributed a quarter. As we increased our capitalisation, they maintained their quarter share until the time the company was listed publicly. Still, they have a large share in our business' (Editors of Common Wealth 1999, p. 40).

More dramatic is the recent case of Wang Wenxiang, director of Hong-ren Corporation. Wang is the oldest son of one of Taiwan's most influential tycoons, Wang Yongching, who is president of Formosa Plastics Corporation and Nan Ya Plastic Corporation, both ranked within the top 20 largest overseas Chinese businesses in Asia, in 1999, by the *Yazhou Zhoukan*[3]. Wang had some personal run-ins with his father and decided to quit his top job at Nan Ya. He went to Berkeley to teach for a year and returned to Taipei four years ago to start his own enterprise. He too credited to personal connections and his social network his present success, with six factories running solidly in Guangzhou and two in Shanghai, manufacturing synthetic leather, PVC, BOPP plastic films and industrial fibres and fabrics. He claims to produce 70% of artificial Christmas trees in the world and is planning to step into a joint venture with one of the world's largest PVC fabric producers – Germany's Klockner.

> When I was working under a large organisation, I never had to worry about money. Yet this was the first and most difficulty problem I had to confront when I decided to set up my own business.... Normura Securities [from Japan] was the first who was willing to invest on my business. I had nothing to show to them except my proposal. They looked at it and decided to invest in me. Others would not follow had it not because of them. I am very grateful to them – they were like offering me coal to burn to keep me warm in deep winter.... But I am even more fortunate to have colleagues who used to work with me at Nan Ya willing to join us and said to me, 'I'm with you'.....They followed me not because they had experienced political pressure from Nan Ya. I think there is a certain emotional tie between us, like brothers. The affection is developed through tens of years of working together, as good friends (Wu 1999, p. 79-81).

It was no surprise to find that well rooted networks that were developed organically through patience, trust and time were typically activated and reactivated in these cases to respond to new incentives and new opportunities. Yet close, reliable networking could also work differently in a less admirable way. Tam (1999) commented that the 'dark side' of social networks should be recognised. What happened to the businesses in Taichung as we described in the beginning of this chapter is instructive.

With hindsight, the reason why a regional crisis impacted so drastically on these particular companies owes more to their own behaviour than to the sector in which they were operating. Taichung Jing-Ji (TJJ), for example, was in financial trouble because, for over two years, it was expanding too fast and was reliant too heavily on easy credit from the banks (mortgaging its own stocks to obtain cash, and using the cash to set up an investment branch to buy back its own shares). When smaller companies in Taiwan started to experience financial difficulties in mid-1998, and rumors were circulated that TJJ was the next to go, its share prices began to slip rapidly and without pause so that in three days, the company was close to collapse. The case of Tsai of HLE was typically about greed. He borrowed heavily to invest in shares for quick profit, but he was never ready for the rapid decline of the share prices.

Similarly TCB's demise was the result of its parent consortium's (Guang-San, GS) heavy involvement in property and financial market development. In 1995 GS successfully transformed the city centre of Taichung by building the Sogo Department Store. At the end of the year, it acquired a financial company Shun Yu Tai (SYT). In the next two years, based on the success of GS's Sogo, SYT shares soared. When the Asian crisis started to hit Taiwan, SYT share prices, without exception, began to plunge and it intended to borrow heavily and illegally from TCB to keep its share prices stabilised. However, the plan was leaked to the media and it immediately spurred a run on the bank.

In the case of TJJ, according to an interview with Huang Minghe (CEO), the reason for its rapid expansion into setting up its financial investment arm was because of his network:

> About three years ago, when we saw many companies in Taipei making a killing on the financial market, we thought we could learn from them. We started ours first. We had fair successes and we set up a second, then a third. When we [entrepreneurs] met in Taichung, we also tried to learn from each other new tricks to manipulate the market. We heard stories about how a particular financial company did this and that. The next thing we knew was that they were coming to us suggesting we should play along and they would provide easy credit....Our original intention was to use this to accumulate bigger capital to expand our production in industrial equipment, but once we had successes in the finance market, we couldn't stop....Unfortunately, before we had a chance to re-orient our

direction, we were confronted with the market collapse (Shen 1999, p. 82).

In the case of Tsai, what is interesting is not what he did wrong. The revelation how he could embezzle such a huge amount was striking because he had brought in his high school classmate to be his special assistant, responsible for fixing up things after he made major decisions for the bank without consulting or informing the bank's finance section. As for TCB's demise, it also illustrates that a particular vulnerability of the share market in Taiwan and its relationship to network capitalism lies in the fact that business failure is personal. As Chan (1999, p. 67) explained,

> In Central Taiwan, price fixing of certain shares is particularly strong. One of the major reasons is because people from the central region do not want to lose 'face'. The share prices of their company are indicators of their social status, as a financial manager observed…. This is because personal networks in central Taiwan are much more closely knitted than in the north. Many shareholders are your next door neighbours. When share prices start to drop, the owner of the company somehow feels personally responsible and is ashamed. We heard the story that the mother of the owner of a publicly listed company actually had to buy back from her friends all the shares they had bought when they complained to her bitterly how the share prices had dropped….

Seen in this context, it further aggravated the vulnerability and volatility of the share market in Taiwan, as deliberate manipulation was not only considered justified but as an obligation to be carried out at any cost, in order to save 'face'. For many entrepreneurs, there are no purely business decisions – all business decisions become personal. Worse still peer pressures from other members in the network intentionally or unintentionally compelled each other to 'conform' amongst themselves.

> You'll have to do what the others do in your network if you want to remain a part of it. It's a real temptation when you see others making quick, easy profit from speculation and they are showing you how to profit similarly. You almost feel you owe them a favour, and you'll lose face if you don't succeed[4]….

We have hypothesised elsewhere that perhaps in the recent times of frenetic economic activity, networks may have been stretched and extended and, where the profit incentive of new opportunities was strong, the process of establishing trust has been accelerated. New networks may have been constructed or extended too rapidly as they become too instrumental and opportunistic, so that ultimately they undermined their own control mechanisms and in turn, legitimacy and moral order. Yet the cases we have reviewed here have shown that things are much more complicated than what researchers have been arguing – one is never certain when and under what conditions social embeddedness can be clearly separated from an organisation or a business. In these cases, they were fused intimately.

Zhang and Tam (1999) have argued that social ties or networks are most notably significant at the start up stage of a business. Thereafter, it is the business' reputation for quality and capability that are key to organisational survival. Yet the rise and the continual power and influence of the *Tainan Bang* (Xie 1999) suggests otherwise. More interestingly, as Zhang (1999) has asserted elsewhere, the development of 'production networks' amongst Taiwan's businesses were testament to their maturity. These networks did not rely solely on existing, established networks characterised by social closure or exclusiveness. They were *ex-post* networks that were cultivated, constructed and extended based on rationality rather than emotive *guanxi* (connections).

Her argument may well have won the day. In the current period when global markets have become severely competitive and when the shadow of the Asian crisis is still far from disappearing, many enterprises that have concentrated on manufacturing or technology are strengthening their production networks to make advances. In central Taiwan, their strategic response to the current meltdown is the strengthening of their 'integrated production network' and a further specialisation in production (Chen 1998).

The term 'integrated production network' refers to a grouping of compatible factories in a region, each specialising in the production of part of a manufactured product (Hsing 1998, chapter 2; Lever-Tracy et al 1996, p. 140). For example, the production of mobile phones requires various components produced by other factories. The advantage of having these factories all located near one another is that they can form an integrated network that supports one another and helps to cut transportation and production costs while also saving valuable time in meeting export deadlines. As the commodity chains begin to mature, the manufacturing of parts also becomes more specialised. Many factories that used to produce a number of parts have now been upgraded technologically to specialise in producing only one item or specialising in one component of the production process. The immediate advantages for them are that in further specialisation, they minimise unnecessary competition with others. They also require less credit to finance their production as the range of their products narrows. In turn, they become more competitive on the international market by cutting costs.

It is a strategy that Taiwanese entrepreneurs have developed and perfected through years of accumulated experience and reflection. As one of the entrepreneurs commented,

> Most of us in Taiwan are small and medium sized enterprises. We used to think we could not compete with the big giants in the US because we had limited resources. We thought we could not undercut the big corporations in prices because we did not have the economy of scale. We felt we could not do any better because we had to compete with each other. What we didn't realise was that we were manufacturing almost exactly the same thing, spending too much time spreading ourselves too thin and we ended up having to spend even more time solving the problems we had created out of competition. Until we realised that there was another way out: co-operation rather than competition[5].

Zhang and Li (1998, p. 10) called this the realisation of *Juzhi Fenjin, Juli Fenjin* (united wisdom brings gold, consolidated efforts bring profits), and it has been producing results not only in Taiwan, but also in mainland China for many Taiwanese funded enterprises. In Taiwan, how the integrated production networks helped to launch the island as the world's major producer of umbrellas, shoes and bicycles is well known (Chen 1994, 1998; Li 1998). Now many take pride in not only seeing the system of integrated production networks firmly transplanted onto mainland China but also how well it worked, enabling them to take advantage of the cheap labour and resources offered by local townships. Gao et al reported with amazement:

> How long did it take to build an empire of bicycle production? In the case of Taiwan, close to 50 years of unrelenting efforts. If this empire were to be reproduced, how long would it take? A mere five years in the townships of Longhua and Buji which the Taiwanese businessmen nicknamed 'Taiwanese bicycle towns' located in Shenzhen....Within half an hour's ride from Shenzhen, on one side of a six-lane expressway... one sees the bright orange neon sign of Meilida, the bicycle manufacturer. Another ten minutes, on Renmin Road in Longhua, you see one after another names that are well known on the international markets: Chiaohui who produces bicycle chains, Liyu (accelerators), Xiangju (paddles), Weikang (seats). Turn around, there appears Xinlong who manufactures handles and bicycle frames, Jianda (tyres), Aiboshi (brakes), Zhengshen

(stickers)...and further down the track, Jinfu and Wufu producing paper cartons and other spare parts. Within a ten-kilometre circumference, over 60 factories are grouped together forming a complete bicycle production linkage system resembling a collection of satellites .... forming a unique mutual assistance system: people borrowing vehicles from their neighbouring companies; if cash flow becomes problematic, a phone call will have it fixed (Gao et al 1997, p. 86).

And this is not the only place in China where such an integrated production network can be found. Gao et al (1997, p. 87) observed that in 1994, in Quanshen, Taicong and Nanxiang, a similar set up was in place involving 20 to 30 Taiwanese manufacturers producing bicycles. More recently, Xijin (1999) reported that in Dongguan in the Pearl River Delta, more than 800 Taiwanese firms similarly had their computer and information technology industry production network operating, producing monitors, hard disks, scanners, keyboards and mouses. Within a circumference of 50 square kilometres, 'one could effortlessly find every single piece of component required for assembling a computer' (Ji 1999, p. 20). An identical development was also replicated in Greater Suzhou (including Suzhou, Wujiang and Quanshen) with a string of 'heavies' in Taiwan's information technology sector (Acer, Philips, Lijie, Chuanyou, Renbao, Guoju, Gaochang, Canquan and Acer's subsidiary, Huashi), rushing in (Wu 1999, p. 188).

Yet the important questions remain – in Taiwan, the integrated production networks are the consequence of a long evolution out of trial and error; but how were these production networks set up in China by the Taiwanese? Particularly significant is that it required an unprecedented 'organised and collective decision to shift all upstream and downstream production simultaneously' from Taiwan to China. This took the Taiwanese government by total surprise and eventually alarmed them with its potential to 'hollow out' Taiwan's own industrial base (Gao et al 1997, p. 87). Are these pure impersonal, rational company decisions, or are they fused with personal preferences? The answers from our informants are instructive:

> There is no clear-cut situation. If you are not the *long-tou* (dragon head), you know, leading manufacturer, you'll have little choice but to make simple company decisions to follow the leader. For example, what good does it do if you're producing parts for a computer if Acer decides to move its assembly to China? You'll have to follow them whether or not you like it. It's pure business unless you have something against them personally. If you are the *long-tou* and you have the power to dictate,

certainly you can be choosy and select just any one to produce the parts for you. But I doubt very much if any business entrepreneur would just pick up a Yellow Pages and randomly choose a manufacturer because you want to make sure the producer is reliable, trustworthy and able to deliver their goods promptly and with good quality. You know you cannot afford to take any risk and have your business stuffed up. So you'll begin with people you know, or ask them for connections and you'll check their reputation and credibility[6].

It is not a matter of just personal preference to pick someone you know to be included in a production network. If this minimises your risks, why not? It's a kind of insurance against uncertainty.... I am not saying that you don't go out of your way to get the right person or the right manufacturer to be part of your production network – you choose them, they also choose you. But you'll need to have a lot in common to continue working together[7]....

What concerned most of our informants however, was that the integrated production network was seen to be crucial not only for the survival of the Taiwanese businesses in the shadows of giant corporations overseas, but also in allowing them to continue to expand, upgrade and develop their operation with a transnational or global ambition, especially when they shift their base of production to China.

We ran into many bottlenecks in Taiwan. Of course, escalating labour costs have been a major problem, but what really crippled us was the fact that few young people in Taiwan now have any interest in working in the manufacturing sector. We begged, pleaded for people to work in our factories, and we were willing to pay a decent wage. But there were no takers. Even if we had all the intentions to expand, to upgrade ourselves, we couldn't[8].

Gao et al (1997, p. 88) illustrated this with a concrete example:

Take the job of soldering bicycle frames.... In Taiwan a manufacturer had to do this by outsourcing, but we couldn't find anyone to do it even when we offered NT\$ 60,000 or 70,000 a month...... Here in China...for NT\$2,000 a month ... you get precious, down-to-earth and hard workers.

Chang Tai, a company producing machinery for carpentry is one of the many who benefited. Its current product was rated top by the Carpentry Magazine in the United States and its sales volume has jumped from NT\$3 billion to NT\$12 billion in three years between 1996 and 1999. It is expecting a 30% growth this year, as it is now producing for Sears and Black and Decker, two of the biggest chains in the US. Ming Yang has also achieved its phenomenal growth in the last six years with a technical breakthrough in producing headlights for motor vehicles. Since then, it has abandoned its traditional lower-end Southeast Asian market and moved into the European market producing for Mercedes-Benz and other brands of vehicle of similar class. It has also established its own line, Depo, for export (Diu 1999).

He-da Industrial, a company producing gear wheels, is another case which benefited from the network that allowed them to have a technological upgrade, and has successfully penetrated both the American and European markets. It is now producing for Chrysler and Triumph, amongst others. It presently employs 240 workers but has a sales volume of NT$700 million with a profit of over NT$90 million. Similarly Quan-Tong has become the world's largest manufacturer of brakes for bicycles by abandoning most of its domestic market and shifting to export. Currently the world's top mountain-bicycle, 'Mounty', uses its Quando line brakes. Its sales volume in the last three years has averaged over NT$600 million annually (Diu 1999).

Shengbao, another manufacturer of electrical appliances went one step further with plans to invite mainland Chinese partners from Shanghai and Tianjin to join its production network, taking on design and engineering tasks. The owner, Chen Shengtian insisted that China, like Taiwan, has an equally solid reserve of engineers with strong technological know-how and one should take advantage of its talents for technological innovation (Qin 1998, p. 92).

This description of the networked small firms of Taiwan is reminiscent of accounts of the 'Third Italy' (Pyke *et al*, eds.,1990; Storper and Scott, eds., 1992). Bagnasco first coined the term, in 1977, to identify some 60 districts that had experienced recent growth of modern industrial firms, in northeast and north central Italy. He contrasted it with the 'First Italy' of heavy industry and auto production in the north west and the stagnant backward rural areas of the 'Second Italy' in the South (Brusco, 1990, p. 16). These districts were local areas, which contained growing clusters of mainly small and also medium firms, often in small town and rural areas which had experienced widespread recent industrialization since the 1960s.

Theorists of 'post Fordism' and 'flexible specialisation' attributed the growth here of dense, vertically disintegrated networks of small firms to a more general shift toward flexible production and away from high-volume standardized production (Piore and Sabel, 1984; Harvey, 1989). These firms made small batches of specialized goods for fast-changing niche markets with a high levels of product differentiation, and a heavy reliance on innovation and quick supply and delivery.

In the new paradigm these districts of small firms flourished, it was argued, through their ability to be innovative and flexible in the new period which privileged such qualities. Harrison (1992) argues that rather than simply the reduction of transaction costs and the realization of external economies of scale, which are emphasized by the 'neo-institutionalists' (Williamson 1975; 1985), the primary advantage of these spatially concentrated firms was the embedding of transactions and contract in long-term trust relationships. Moreover, what enabled them to avoid both the sclerosis of mass production and the weakness and dependency of small firms, was that they operated as 'regimes of cooperative competition', combining entrepreneurial rivalry with cooperation and trust, both of which were intensified by the spatial concentration of designers, suppliers, and producers (Capecchi, 1990; Harrison 1992; Lorenz 1992).

During the 1970s and 1980s their production was growing fast (against the current of a period of national industrial decline), with increasingly sophisticated products, skills and use of high technology leading to rising wages and living standards. The per capita income of Emilia Romagna, for example, rose from 17th out of 21 Italian regions in 1973 to second in 1986, with wages at twice the national average (Harrison, 1992, p. 472). Many of these firms sold self designed products, directly to markets around the world. Locally based export agents, the *impannatori,* followed world market trends and worked to relate local design to global tastes and to translate latent capabilities into saleable products (Becattini, 1990, p. 42).

Conclusion

In the beginning of this chapter, we observed that for quite some time researchers have been debating whether social networks and their 'embeddedness' are both valid and useful in accounting for the successes of diaspora Chinese enterprises, as conventional wisdom has claimed. On the contrary, some argue that when businesses mature, the importance of personalised networks will be replaced by organisational cooperation and interests. Implicit in this debate is the suggestion that networks and their embedded social relations could be corrupting, promoting personal greed, debt, over-expansion of business and ultimately economic bubbles, particularly in times of frenzied economic activities that promise new quick profits and glamour.

Our ongoing research is examining a range of case materials and secondary sources on Taiwanese businesses and entrepreneurs, to document how at the present time social networks are being formed, used and developed at various stages of the business cycle. At this stage we can only propose some preliminary impressions to guide further research, and cannot claim to resolve the debate between the social embeddedness and the organisational imperative theses. Since Western observers have been highly critical about the use of networks among diaspora Chinese entrepreneurs during the Asian crisis, we are also interested in finding out whether such social networks still occupy the same sense of importance in their businesses and business lives. We are particularly interested in the indications we have found that such social embeddedness has helped them to survive, consolidate or expand in times of rapid change.

After reviewing the case materials, we could only arrive at some tentative conclusions. There was little evidence to indicate that the use of social networks among Taiwanese entrepreneurs and their businesses is necessarily uniquely immoral. In the cases we reviewed, most used their long tested and trusted networks as social, if not always economic capital, to strengthen their business' capacity to survive. The Tainan Gang developed their close-knit and complicated networks to demonstrate that they were a force to reckon with from the south of Taiwan. Tainan Enterprises rebuilt their business from the brink of closure to success by relying on the strength they drew from their own family members. Quanta would not have been in business had it not received the timely capital injection from some key old colleagues and friends. And the eldest son of one of Taiwan's most influential tycoons wouldn't have been able to reinvent himself in the business world without his old network of colleagues and clients from Japan. The series of crashes of businesses in central Taiwan in 1998 were reminders of how social networks could be corrupted, but the events revealed more about the actors own behaviour rather than any inherent immorality of networking. If anything, one could even argue that social networks, despite their seemingly exclusive and secretive facade, are in fact highly moralistic. The fact that the offender in the economic scandals in Taichung admitted his wrong-doings publicly in the end was indicative of his sense of shame for abusing what is considered sacred in a network – trust. In using networks to enhance the chances for their business to survive and to remain competitive, the nature of networks in Taiwanese business did evolve into something broader than personal.

The emergence of production networks in many ways are confirmation of the view represented by the organisation imperative thesis – when businesses mature, cooperation between and among businesses will necessitate the creation of new and expansion of old networks. The eventual appearance of the well-tested integrated production network could also be seen as the inevitable outcome of organisational imperatives. However, this could also be deceptive. For every new network that is cultivated, constructed or expanded also involves a personal choice. Our informants have suggested that the preference has always started with a choice that one is more familiar with – embedded in existing social relations – rather than one that inherits strangeness and uncertainty. In this context, the social embeddedness thesis also makes sense. Given the limited information we had, we do not feel confident to arrive at a final resolution to the debate.

The production network, and ultimately the integrated production network, seems to have become a most effective strategy for Taiwan's businesses not only to survive and to grow but also to transform themselves in the new world of globalisation. Their ability to replicate a well-tested system of integrated production network in various regions in mainland China perhaps is crucial in accounting for the rapid prominence and confidence some Taiwanese business displayed on the international stage, particularly in proclaiming their ambitions to make a giant leap from sunset industries to sunrise industries, from ODM to OBM, or even JDM[9] (CommonWeath Editors 1999, Qin 1998).

The biggest doubts in discussions of the 'Third Italy' were first about the extent to which it was a special case that could not be generalised and was unlikely to spread beyond Italy. The second was about its capacity to survive, with any degree of autonomy, in a global world supposedly ruled by multinationals (Amin and Thrift, 1992). The Taiwan case tends to offer some reassurance on both these doubts.

Notes

[1] Typically they include Wu Juan-xian, Wu Xiu-chi, Gao Qing-yuan, Hou Bo-yi, Hou Bo-yu, Hou Bo-ming and others, representing businesses such as the Uni-President Enterprises Corporation, President Chain Store Corporation, Tainan Spining Co. Ltd. and Universal Cement Corporation. See Xie 1999, chapter 4 and 5.

[2] *Yazhou Zhoukan The International Chinese Weekly*, November 1-7, 1999, pp. 34-35.

[3] *Yazhou Zhoukan The International Chinese Weekly*, November 1-7, 1999, pp.34-35.

4   Personal interview with informant on November 24, 1999 in Taipei as part of an on-
    going research on Responses of Diaspora Chinese to the Asian Crisis funded by the
    Australian Research Council 1999-2000.
5   Author's interview in Taipei, 24 November 1999.
6   Interview with informant, 26 November 1999, Taipei.
7   Interview with informant, 23 November 1999, Taipei.
8   Interview with informant, 23 November 1999, Taipei.
9   Stands for Joint Design Manufacturing. Quanta now involves JDM with Apple
    computers. See Commonwealth Editors 1999, p. 39.

References

Amin, A. and Thrift, N. (1992), 'Neo-Marshallian Nodes in Global Networks', *International Journal of Urban and Regional Research*, 16.

Becattini, G. (1990), 'The Marshallian Industrial District as a Socio-Economic Notion' in Pyke, F., Becattini, G. and Sengenberger, W. (eds) (1990), *Industrial Districts and Inter-Firm Cooperation in Italy*, IILS, Geneva.

Brusco, S. (1990), 'The Idea of the Industrial District: Its Genesis', in Pyke, F., Becattini, G. and Sengenberger, W. (eds) (1990), *Industrial Districts and Inter-Firm Cooperation in Italy*, IILS, Geneva.

Capecci, V. (1990), 'A History of Flexible Specialisation and Industrial Districts in Emilia Romagna', in Pyke, F., Becattini, G. and Sengenberger, W. (eds) (1990), *Industrial Districts and Inter-Firm Cooperation in Italy*, IILS, Geneva.

Chen, J.S. (1994), *Xueli Wangluo Yu Shenghuo Jiegou – Taiwan Zhongxiaoxing Qiye De Shehui Jingji Fenxi* (Cooperative Networks and the Structure of Living – An Economic Analysis of Small and Medium Size Enterprises in Taiwan), Lian-jing Publishing Co., Taipei.

Chen, J.S. (1998), *Taiwan Chanye de Shehuixue Yanjiu -- zhuanxing de zhongxiao qiye (Sociological Study of Taiwan's Assets -- Middle and Small Size Enterprises in Transition)*, Lianjing Publishers, Taipei.

Chen, Y.W. and Chan, Y.S. (1999), 'Yinhang Lauban Qiang Yinhang (Bank Boss Robs Banks?)', *CommonWealth*, April, pp. 76-79.

Christerson, B. and Lever-Tracy, C. (1997), 'The Third China? Emerging Industrial Districts in Rural China', *Urban and Regional Development*, December.

Diu, M.P. (1999), 'Buchao Gupiao Ye You Gao Chengzhang (High Growth Without Playing the Stock Market)', *CommonWealth*, April, pp. 86-89.

Editors of CommonWealth (1999), *Xin Jingji Xin Jihui Xin Lingxiu* (New Economy New Opportunities New Leaders), *CommonWealth Magazine*, Taipei.

Gao, X. (Kao, H.C.) et al (1997), *Taisheng Jingyan -- Touzi Dalu De Xianchang Baodao* (The Taiwan Investment Experience in Mainland China -- A First-hand Report), Commonwealth Publishing Co. Ltd., Taipei.

Granovetter, M. (1985), 'Economic Action and Social Structure: The Problem of Embeddedness', *American Journal of Sociology*, Vol. 91, pp. 481-510.

Guan, Z.S. (1999), 'Yizhen Yixian Zhichu Xiwang Chanye' (Weaving a Promising Enterprise Needle by Needle and Thread by Thread), *CommonWealth*, November, pp. 60-66.

Harrison, B. (1992), 'Industrial Districts: Old Wine in New Bottles?', *Regional Studies*, 25/5.

Harvey, D. (1989), *The Condition of Postmodernit*, Basil Blackwell, Oxford.

Hsing, Y.T. (1998), *Making Capitalism in China: The Taiwan Connection*, Oxford University Press, New York and Oxford.

Ji, S.M. (1999), 'Dongguan Moshi Liangan Sanyin' (Dongguan Model, 3 Winners for Both Sides), *Yazhou Zhoukan The International Chinese Weekly*, October 18-24, pp. 18-22.

Lever-Tracy, C., Ip, D. and Tracy, N. (1996), *The Chinese Diaspora and Mainland China: An Emerging Economic Synergy*, Macmillan, Houndmills.

Lorenz, E. (1992), 'Trust, Community and Cooperation. Towards a Theory of Industrial Districts', in Storper, M. and Scott, A.J. (eds), *Pathways to Industrialisation and Regional Development*, Routledge, New York.

Numazaki, I. (1993), 'The Tainanbang: The Rise and Growth of a Banana-Bunch-Shaped Business Group in Taiwan', *The Developing Economics*, December, pp. 503-510.

Pyke, F., Becattini, G. and Sengenberger, W. (eds) (1990), *Industrial Districts and Inter-Firm Cooperation in Italy*, IILS.

Qin, M. (1998), *Kaichuang Taiwan Dianzichuan Renwu Fengyun* (The Pioneers of Taiwan's Electronics Industry), Daotian Publishing Co., Taipei.

Shen, J.X. (1999), 'Ruguo Neng Chungxin Zhailai' (If I Could Start All Over Again), *CommonWealth*, April, pp. 80-85.

Summers, L. H. (1998), 'Opportunities Out of Crises: Lessons From Asia,' remarks by Deputy Treasury Secretary to the Overseas Development Council, statement released from the Office of Public Affairs, March 19.

Tam, T. (1999), 'Corporate Networks in Taiwan: An Overview', in Zhang, L.Y. (ed), *Wanglou Taiwan* (Corporate Networks in Taiwan), Taiwan Industry Research Mook 2, Yuan-liou Publishing Co., Taipei, pp. 285-291.

Williamson, O.E. (1985), *The Economic Istitutions of Capitalism*, The Free Press, New York.

Wu, Y.Y. (1999), 'Wangzhi Fuxing Sannian Buwan' (Never Too Late for the Re-emergence of a Prince in Three Years), *CommonWealth*, November, pp. 78-84.

Xie, G.C. (1999), *Tainan Bang: Yige Taiwan Bentu Qiye Jituan De Xingqi* (The Gang from Southern Taiwan: The Rise of a Taiwan Native Business Group), Yuan-Liou Publishing Co., Taipei.

Zhang, L.Y. (ed) (1999), *Wanglou Taiwan* (Corporate Networks in Taiwan), Taiwan Industry Research Mook 2, Yuan-Liou Publishing Co., Taipei.

Zhang, L.Y. and Tam, T. (1999), 'Xinggu Chanye Wanglou' (Form and Structure of Production Networks) in Zhang, L.Y. (ed) (1999), *Wanglou Taiwan* (Corporate Networks in Taiwan), Taiwan Industry Research Mook 2, Yuan-Liou Publishing Co., Taipei, pp. 17-64.

7 Diaspora Chinese Enterprises in Guangdong – Problems and Government Responses During the Asian Financial Crisis

CEN HUANG

Introduction

In May 1998, three provincial government offices (the Provincial Committee of the Chinese People's Political Consultative Conference, the Provincial Liaison Committee for Compatriots from Overseas, Hong Kong, Macao, and Taiwan, and the Provincial Office of Overseas Chinese Affairs) in Guangdong conducted a series of field investigations in eight cities and counties in the Meizhou and Shantou region where overseas Chinese investments were heavily concentrated. The objective of the investigations was to find out how the Asian financial crisis had affected enterprises funded by overseas Chinese in the province and how the government could help them to overcome the difficulties they had encountered.

Concurrently, local government offices also carried out similar surveys in local small towns and villages. This paper reviews some of the major findings from these government reports. It also summarises the impacts of the Asian crisis as perceived by the entrepreneurs who were interviewed by the author. Case materials from Chaozhou and Shanwei will be used to highlight the impact of the Asian crisis on local regions, and a series of government policies, initiated by the provincial and local governments, to assist overseas Chinese entrepreneurs to overcome the problems they have encountered, will be evaluated.

Impacts of the Asian Financial Crisis on Guangdong Province

Eight cities and counties in Guangdong Province were surveyed by the provincial and local governments. They included Meizhou, Shantou, Jieyang, Shanwei, Chaozhou, Chaoan, Chaoyang, and Chenghai in the Meizhou and Shantou region. Results of these surveys revealed some noticeable impacts of the Asian crisis, that had been inflicted quite heavily on Guangdong Province. Foremost, foreign investments, exports, donations inspired by *qiaoxiang* ties (loyalty of migrants to their place of origin) were negatively affected by the current crisis. Furthermore, complaints from overseas Chinese entrepreneurs, operating in the province, about unfair differential tariff rates, proliferation of fees and charges and bad debts, were also observed with increasing frequency.

Declining Foreign Investments

The amount of foreign investments in Guangdong had declined steadily since the beginning of the Asian crisis in mid 1997. In Shantou, one of the three Special Economic Zones in Guangdong, 62 new projects, worth a total of US$75.49 million, were approved in the first quarter of 1998. Although this meant that the number of new projects went up by 32%, the value of total investments actually had gone down by 46% when compared to the same period in 1997. Moreover, although seven key investment projects (US$30 million each), funded by foreign capital, were secured in 1997, none were actually carried out in 1998 (The Guangdong Provincial Liaison Committee for Compatriots from Overseas, Hong Kong, Macao, and Taiwan, 1998a).

Other statistics also showed that from July to December 1997, despite the fact that 13 new foreign projects (with a total investment of US$31.2 million) were approved in Chaozhou, the total number of projects and amount of investments, had in fact gone down by 53.6% and 47.3% respectively since the same period in 1996 (The Commission for Foreign Economic Relations and Trade, Chaozhou, 1998). Similarly in Jieyang, although 107 foreign funded projects were planned in 1997, the total investment had been reduced by 54.3% from the same period in 1996 (The Guangdong Provincial Office of Overseas Chinese Affairs, 1998a). In Chenghai, 39 projects were approved in 1997 but 33 were finally cancelled because of funding difficulties in 1998 (The Guangdong Provincial office of the Overseas Chinese Affairs, 1998a). Reports also revealed that in the first quarter of 1998, Japanese investments in Guangdong were down 42%, while investments from South Korea concurrently decreased by 50% compared to the same period in 1997 (*The South China News*, 13 July 1998).

Similarly in Meizhou, it was reported that since 1997 investments in the city from Southeast Asian Chinese had come to a complete halt. In the previous year there were 68 (The Commission for Foreign Economic Relations and Trade, Meizhou, 1997) but there was none on the drawing board in 1998. Moreover, many projects commenced earlier were forced to close down as funding evaporated. In Shantou, the PVC Industrial Raw Materials Project, a US$10 million project funded by Hong Kong, Thai and Japanese investors, established in 1996, had constructed its factory buildings and imported production equipment. However when the financial crisis hit Thailand, the Thai partner withdrew immediately. The Japanese and Hong Kong partners were reluctant to continue the project. It is now completely abandoned. Another project, the Shunheheng Rice Production Base, a HK$20 million project funded by Thai Chinese investors met a similar fate (The Guangdong Provincial Liaison Committee for Compatriots from Overseas, Hong Kong, Macao, and Taiwan, 1998a).

All eight counties and cities surveyed were locations traditionally heavily reliant on *qiaoxiang* investments – for example, 90% of foreign investments in Shantou, 85% in Chaozhou, 70% in Zhaoyang, 80% in Jieyang and Chaoan, were all from overseas Chinese, and particularly from Southeast Asian sources. Notably all these locales are now suffering badly under the Asian financial crisis (The Guangdong Provincial office of the Overseas Chinese Affairs, 1998a).

Declining Exports

During the Asian crisis, many countries in the region depreciated their currencies massively. This has posed great difficulties for exports for Guangdong as it found difficulty in competing with other Asian exporters who were asking lower prices while the Chinese *yuan* remained strong. It was reported that orders from traditional export markets (mainly the US and Europe) had declined at an average between 20-30% in Guangdong province in 1997 (The Guangdong Provincial Commission for Foreign Economic Relations and Trade, 1998). Adding salt to injuries, according to reports from Chaozhou, its exports to Southeast Asia went down by 32.2%, to Japan down by 44.6% and to South Korea by 20.9% in the first five months of 1998 (The Commission for Foreign Economic Relations and

Trade, Chaozhou, 1998), presumably because they were still suffering from the fallout of the crisis.

Worse still, international buyers were reportedly demanding lower prices by threatening to switch to other cheaper sources if Guangdong did not comply. Not surprisingly at the Spring Trade Fair of Guangzhou in May 1998, export prices of all commodities were reported to suffer from a 20% reduction when compared the same period in 1997. Export prices went down even further in some local regions. For example, in Shantou, prices for machinery dropped by 50%, and garments, 40% (The Guangdong Provincial Liaison Committee for Compatriots from Overseas, Hong Kong, Macao, and Taiwan, 1998a). In Punning City, export prices for food slipped 25%, while garments fell by 20% and toys 15% (The Commission for Foreign Economic Relations and Trade, Punning, 1998).

Nevertheless, the good news that emerged from the gloom was that imported raw materials, equipment and technology were cheaper due to the strong *yuan*. That explains why imports from Southeast Asia went up by 70% in the first quarter of 1998 in Chaozhou (The Commission for Foreign Economic Relations and Trade, Chaozhou, 1998).

Reduction of Qiaoxiang Donations

Overseas Chinese have always played an important role in Guangdong's development through their generous donation, either in monetary terms or in terms of improvement of social and physical infrastructure such as building schools, hospitals, roads, or instituting new welfare services. In the past two decades, almost every *qiaoxiang* village in Guangdong has benefited from such *qiaoxiang* donations (Huang, 1998). However, these donations have been sharply reduced since 1997. A survey found that in the past 10 years Chaozan County benefited by as much as 30 million *yuan* annually from *qiaoxiang* donations. In the first six months of 1998, however, only 3 million *yuan* were received. Consequently many welfare projects in the county had to be discontinued (The Guangdong Provincial office of the Overseas Chinese Affairs, 1998a). Significantly reduced also were overseas Chinese visitors to Guangdong in 1997 and 1998. The China Travel Agency (Shantou) reported that the number of Southeast Asian Chinese visited Shantou declined by 60% in the first quarter of 1998 (The Guangdong Provincial office of the Overseas Chinese Affairs, 1998a).

Complaints of Overseas Chinese Entrepreneurs

Confronted by the fallout from the crisis, many overseas Chinese entrepreneurs also found themselves experiencing difficulties that may not have been new but which now had inflated to a proportion that would affect the viability of their businesses. These problems included:

Non-standardised tariff rates Tariffs were imposed on overseas Chinese businesses in Guangdong Province for importing non tax exempted materials and for applying to sell their products in the domestic market. However, because of the uneven development of rural industrialisation, different tariff rates had been set up by different local customs offices to attract potential investors. Each year each local customs office has its own obligations for filling a quota of tax revenues as directed by the provincial government. In meeting the quota, local offices thus take advantage of the flexibility they have in setting their own tariff rates. In times of prosperity, this would not have become a real problem for business owners.

However, since the beginning of the crisis, the provincial government had to maintain its own revenue from taxes. In turn, there had been increasing pressures on local customs offices to fill their tax revenue quota. Thus, for example, when the quota for the Chaoyang customs office imposed by the Shaotou Regional Office increased from 40 million in 1996 to 52 million in 1997, and finally 60 million *yuan* in 1998 (The Guangdong Provincial Liaison Committee for Compatriots from Overseas, Hong Kong, Macao, and Taiwan, 1998a), they had to pass on the increases to the diaspora Chinese businesses. Yet since only few imports or exports went through the Chaoyang customs office, the increases were at such an astronomic rate that many overseas Chinese investors felt compelled to complain as they were also facing credit problems from the Asian crisis. Some chose to relocate their enterprises in other counties where the tariff rates were more favourable, such as those in Dongguan.

Proliferation of fees and charges Although the provincial government stated that in the face of the crisis, it was their intention to reduce taxes and fees on foreign-invested enterprises, many overseas Chinese entrepreneurs felt that there were still plenty of unjustified fees and charges. In Jieyang in Shantou Region, it was found that foreign enterprises had to pay more than 40 items of fees and charges that included public security, public hygiene, social welfare, labour bureau, environment protection, fire prevention, birth control, customs, administration and others. Excessive fees and charges were reported as one of the major reasons why close to half of the 250 foreign-invested enterprises chose to stop their production in early 1998 while another 60 decided to fold by the end of 1997 in the Rongcheng District in Jieyang (The Guangdong Provincial office of the Overseas Chinese Affairs, 1998a).

A closer examination of the structure of the fees illustrates well their excessiveness. For example, electricity and water charges were 200% higher in the Shantou Region than in the Pearl River Delta Region. The shipping cost from Shantou to Japan was US$1200 per container but only US$400 if shipped from Xiamen (The Guangdong Provincial office of the Overseas Chinese Affairs, 1998a). It was not uncommon for many to move their business operation elsewhere.

Cash flow difficulties Since the beginning of the Asian crisis, it was found that delays in payment of processing fees from international contractors and bad debts from international buyers were common. For example, delays of payment were extended from 30 days to five to six months. This created serious cash flow problems for many manufactures. In Longhu District in Shantou, over 10 enterprises had to fold because of their liquidity problems (The Longhu District Office of the Liaison Committee for Compatriots from Overseas, Hong Kong, Macao, and Taiwan, 1998).

The Big Picture

Despite the problems outlined, Guangdong Province remains a leading exporter and recipient of most foreign investment in China. Its foreign trade and economy continued to grow. For example, in the first quarter of 1998, the number of new projects approved involving foreign capital went up by 16.2% in numbers and 38.6% in value from the previous year. Exports (US$17.6 billion) and actual inflow of foreign capital (US$3.06 billion) also increased in 1997 (*The Guangzhou Daily*, 2 September 1998). Investments from Japan, South Korea, Hong Kong and Southeast Asia declined, but the increased capital flowing from Europe and America more than compensated what it had lost from Asia. Statistics show that during the first six months of 1998, Guangdong received from Europe investments of US$ 1.2 billion (up 60.62 % on the same period in 1997) while American investments were also up 86.91 % (US$ 71.27 million) (*The Guangzhou Daily*, 2 September 1998).

Unexpectedly, investment from Taiwan made a big leap. During the first four months of 1998, Taiwanese investment in China went up by 41.2% (US$550 million) from the same period in the previous year, pushing Taiwan's total investment in China at the end of June 1998 to US$19.16 billion (*The South China News*, 13 July 1998). Guandong's gain, however, was at the expense of other Asian countries. As the Asian crisis deepened, investments from Taiwan in Indonesia went down by 96.6%. A similar trend was observed in the Philippines (down 87.69%) and Thailand (down 45.7%) (*The China Times* 2 September 1998).

The major beneficiary however was Dongguan. At the end of 1997 more than 2400 Taiwanese firms were established in the city. In the first six months of 1998, more than 300 new projects from Taiwan were approved, taking the actual investment from Taiwan in Dongguan to a total of US$ 2.4 billion at the end of June 1998 (*The Guangzhou Daily*, 2 September 1998). We were told by several Taiwanese investors there that they were planning to re-invest 30% of their annual profit and expand new production lines, as well as finding new business partners to join them from Taiwan. Recent reports from Shenzhen also indicated a strong Taiwanese interest – contracts for 27 projects worth US$404 million from Taiwanese investors were also signed during the Guangdong-Taiwan Economic, Technology and Trade Fair in April 1999 (*The Guangzhou Daily*, April 18, 1999).

Case Materials from Chaozhou and Shanwei

In spite of the more upbeat picture Guangdong Province displayed, some areas in the province were experiencing more difficulties than others. In this section, we will examine the specific impacts on the overseas Chinese enterprises in Chaozhou and Shanwei.

Chaozhou

Of all the 1,082 foreign-invested enterprises in Chaozhou City, 932 (86.1%) were funded by capital from Hong Kong, Taiwan and Macao, while only 106 (9.8%) were from Southeast Asian, Japanese and South Korean sources. Most of these enterprises were export-oriented manufacturing firms. About 80% of the finished products were exported through Hong Kong with only 9% directly exported to Southeast Asia. In 1997, foreign investment in Chaozhou amounted to US$164 million, down

31.8% from 1996. The impact of the economic downturn was highly visible throughout the city:

- For example, Sanyao Porcelain Manufacturer (a Hong Kong enterprise) had to lay off half of its workforce because of the sudden loss of its major clients in Malaysia and Indonesia due to their economic crisis.
- Sanghuan Electronics Inc. (a Taiwanese company) was forced to reduce the prices of its products (electric resistors) by 10-15% between October 1997 and April 1998. Its original projection for 1998 was a surplus of 485 million *yuan*. It had suffered a loss of 20 million *yuan* as the year ended.
- Jinman Group, a food processing company owned by Hong Kong and Southeast Asian Chinese, specialising in grilled eels for the Japanese market, had to cut back 20% of its export. The prices of its product also suffered a drop of 17.6% in 1997. The outcome was a deficit of 20 million *yuan* because of the downturn of the market.
- Jinlibao Textiles (owned by Southeast Asian investors) also found itself in financial trouble as the Malaysian currency was losing ground. Since its major market was in Southeast Asia, all accounts were settled in Malaysian currency. The depreciation of the currency meant big increases for importing raw materials and bad debts from clients in the region. Consequently, one-third of its production lines were halted (The Commission for Foreign Economic Relations and Trade, Chaozhou, 1998).

Shanwei

There were 839 foreign-invested enterprises in Shanwei in 1997. All were owned by overseas Chinese. In 1997 actual foreign investments in the city amounted to US$1.53 billion, but more than half of the enterprises in the city suffered big blows from the Asian financial crisis:

- Xinliang, a Taiwanese manufacturing firm specialising in calculators, had a workforce of 280 workers. It exported its products 100% to Japan. However, because of the crisis, the market was down by 50% and the company had to operate at half of its capacity.
- Qiaoye Electronics, another Taiwanese enterprise started in 1992, aiming at the South Korean and Japanese market. In 1997, demand for its product suffered a 70% drop. It folded in early 1998.

- Yajing Garments, a Hong Kong company established in 1989, with its market in Southeast Asia found itself receiving no order from its traditional buyer all year in 1997 and during the first six months of 1998. It went out of business in 1998 and all 230 workers were unemployed.
- Similarly, Huayi Agricultural Products, a Southeast Asian Chinese company started in 1993, suffered from a loss of market demand after the Asian crisis struck. There were no orders from its traditional Southeast Asian clients and it closed down in 1998.
- Jiahua Garments was another Hong Kong Chinese enterprise, begun in 1990, specialising in hats and bags for the Southeast Asia market. The company also went out of business as the market disappeared in late 1997.
- Fengshan Shoe Manufacturer, a Taiwanese company exporting to Japan and the Philippines, suffered a 15% drop on its export orders in 1997 and 20% of its workers were retrenched.
- Xinxing Electronics was a Hong Kong company with a 100% Japan client-base. In 1997 its business went down by 30% and the company was half-closed.
- Wansheng Garments, a Hong Kong company, similarly lost HK$2 million on exporting to South Korea in 1997, because of the dwindling demand and massive depreciation of the Won. One-third of its workers were sacked (The Commission for Foreign Economic Relations and Trade, Shanwei, 1998).

Reports from Overseas Chinese Entrepreneurs

We talked to 14 overseas Chinese entrepreneurs in the four cities of Guangzhou, Zhongshang, Dongguan and Qingyuan in August 1998, during a short field visit. These informants were selected through personal introduction from local sources. They included five Taiwanese, six from Hong Kong, and three Southeast Asian Chinese. Five specialised in garments, one in agriculture, four in computer technology, three in shoe manufacturing, and one in plastics.

The impressions we gathered from the discussions we had with them were largely consistent with the picture we portrayed earlier. The impacts of the Asian crisis on them were felt unevenly. For example, those who engaged in the computer business were least affected while those in garments found it hard to be immune from the fallout. Those who exported to East and Southeast Asian region experienced a sudden loss of markets, but those with an export base in the American or European continents seemed unaffected. Many said that they were under pressure to lower their prices in order to retain traditional markets, but because of the widespread cut backs in production, labour shortages in the region seemed to be a thing of the past.

Among those who persisted through hardship and losses, some did not feel that business failures were entirely the fault of the Asian crisis. Bad management was cited as more devastating than the crisis itself 'because it had laid a bad foundation for the business', and 'it is those who have good managerial practice and financial management' and 'those who employed professional managers and technicians' who would survive because 'they would be more capable to deal with crisis and risks.'

Few, however, indicated their willingness to switch their base of operation from China. Many believed that the crisis was a temporary setback, and all had their eyes on expanding into China's domestic market at the end of it. Nevertheless, many felt that a diversification of markets was necessary in the future in order to strengthen and consolidate their business. 'South America, Eastern Europe, the African continent are the new frontiers' as one of the informants explained. At least one admitted that he had made considerable profits from textiles when he shifted his market from Indonesia and Malaysia to Vietnam in late 1997. Another shoe manufacturer indicated sales to South America and Mexico alone were his new target.

Notwithstanding, many felt a reduction of production costs and an upgrade in technology would also be necessary for them to meet the challenge of the current crisis. They, however, believed that in the short-run, assistance from local governments would be significant for them to overcome some of the major difficulties they presently encountered.

Government Policies and Assistance

Since early 1998, the provincial, city and county governments have developed specific policies aimed at assisting foreign-invested enterprises to overcome problems arising from the Asian financial crisis. These policies targeted taxation (corporate and personal income tax of entrepreneurs), duties and excise (land lease, customs and export). In some township and village-level governments, more favourable tax systems were planned to attract investments.

Provincial Policies: Assisting Exports

Eight key measures were initiated by the Guangdong provincial government (Wang Xiangwei, 1998):

- Full corporate profit-tax rebates would be offered to state enterprises that increase their export by more than 15% this year.
- Corporate income tax for exporters would be standardised at 33%.
- Subsidised loans would be offered by local authorities to enterprises with a 15% growth in export.
- Full tax rebates on exports would be paid by the provincial government, but from the budgets of local authority.
- Procedures for export companies to be listed on the domestic stock market for capitalisation would be simplified and facilitated.
- Value-added tax and fees on processed goods for export would be waived.
- Local authorities in Zhuhai, Shantou, Shenzhen and Guangzhou would be asked to extend certain favourable central government policies, previously restricted to Chinese export companies, to foreign firms.
- Foreign trade would be further opened to private firms.

Municipal Policies: Improving Investment Climates

The Guangzhou Municipal Government in December 1998 targeted the following main areas to assist foreign-invested enterprises during the crisis:

- All unnecessary fee charges on foreign enterprises would be removed.
- A standardised system of taxes and fees would be applied to all enterprises.
- Procedures for application to set up business in the city by foreign investors would be facilitated.
- Business services offered to foreign investors in the city would be improved.

Local Policies: Tax Reforms in Meizhou

As a response to the Asian crisis and to strengthen investment from *qiaoxiang*, the Meizhou Minicipal Government instituted a series of reform policies in January 1998 (The Meizhou City Government, 1998):

- The rate of corporate income tax for foreign-invested enterprises was reduced from 30% to 24%.
- Corporate income tax would only be imposed on enterprises that intended to operate in China longer than 10 years after they had made a profit consecutively in two years. This policy is called 'two years of exemption and three years of tax reduction by half'.
- A 40% tax rebate would be given to enterprises that had operated for more than five years and for reinvesting its profits in production. A full tax rebate (100%) would be offered if the enterprises were export oriented or engaged in high technology.
- Export-oriented enterprises would continue to enjoy a 50% reduction of corporate income tax after 2 years of tax exemption and three years of tax reduction by half.
- High technology enterprises would be entitled to a 50% reduction in corporate income tax for the next three years after the two years of tax exemption and three years of tax reduction by half.
- No local taxes should be imposed on foreign-invested enterprises located within Meizhou City.
- Foreign-invested enterprises that suffered losses would be allowed to retain the profits they made in the following five years.
- Investors of foreign enterprises were not required to pay personal income taxes on profits made in their enterprise.
- No import or value-added tax to be imposed on selected raw materials and equipment imported when they were used for export production.
- No value-added tax or processing fees would be imposed on export processing enterprises.
- No value-added taxes would be imposed on foreign-invested enterprises engaged in plantation, forestry and animal husbandry.
- Foreign-invested enterprises would be allowed to gradually increase sales their products in the domestic market.

Other Policies: Credit, Domestic Markets and Qiaoxiang Donations

The Guangdong Provincial Liaison Committee for Compatriots from Overseas, Hong Kong, Macao, and Taiwan suggested the following to the Provincial Government to assist those overseas Chinese projects which had experienced financial difficulties. They included (The Guangdong Provincial Office of the Overseas Chinese Affairs, 1998a):

- Foreign investors with good credit records should be allowed to apply for bank loans from banking institutions in China.
- Full export-oriented foreign enterprises should be allowed to sell their products in China's domestic markets and the sales taxes should be waived.
- The government should lift export quotas for economically underdeveloped areas in the province in order to attract foreign investment (The Guangdong Provincial Liaison Committee for Compatriots from Overseas, Hong Kong, Macao, and Taiwan, 1998b).

As a gesture to be more sensitive to the financial situation of overseas Chinese in Southeast Asia, the Office of the Overseas Chinese Affairs decided to stop publicising news concerning investments and donations from Southeast Asian Chinese. It also suggested that requests for donations from overseas Chinese should be halted temporarily. Instead, it would be more effective for large business delegations consisting of state-owned enterprises and private entrepreneurs to be organised to visit communities of ethnic Chinese in Southeast Asian countries to promote trade and investment opportunities.

Conclusion

The Asian financial crisis caught many diaspora Chinese entrepreneurs by surprise. Guandong Province, as a whole, fared very well by all standards when compared to those regions in Asia that suffered massive cutbacks, devaluation of currencies and unemployment. Yet it was also apparent that in specific areas in the province, and for specific enterprises, the fallout from the crisis was less than kind and in some cases, severe. This paper does not aim to dissect, rather it aims to describe how the government, the entrepreneurs and the businesses responded. Given the limitations of time and resources, the interviews we had with our informants were impressionistic rather than systematic.

To summarise what we observed, for those who had their clients based in the East or Southeast Asia region, and for those who engaged in some sectors other than information or high technology, the impacts from the crisis have been serious, if not fatal. The double blow of losses of traditional markets and confronting an archaic as well as chaotic system of excessive excises and fees in China made it difficult for many to survive. However, without any feeling of being heartless, it should be noted that many of those who failed perhaps were in a precarious position even before the crisis struck, as problems of liquidity are frequently long-term rather than sudden. Impressions from our discussions with the entrepreneur-informants certainly have reinforced such an observation. In this context, perhaps what they suggested – professional management and upgrade of technology – were the keys to the future success of diaspora Chinese businesses rather than assistance from the provincial or local governments in China. After all, for both Taiwanese and Hong Kong entrepreneurs, self-reliance has always been the key.

Some entrepreneurs had the impression that government services, taxation and bureaucratic procedures seemed to have improved and simplified during the crisis, but none would want to comment in specific terms. It was reported in early 1999 that a total of 337 items of administrative fees and other charges had been removed or streamlined since the new policies were put in place (*The Guangzhou Daily*, 19 April 1999). Whether the new policies and measures designed to give assistance to overseas Chinese enterprises are achieving their proclaimed objectives, perhaps it is still early to tell.

References

The China Times.

The Chinese Communist Party Committee of Chaoan County (1998), *Dongnaya jinrong fengbao dui wo xian waijingmao gongzuo de yingxiang ji duice* (The Impact of the Southeast Asian Financial Crisis on the Foreign Trade and Economy in Chaoan, and the Policies of How to Deal with It), March.

The City Government of Guangzhou (1998), *Guanyu jinyibu gaishan touzi ruan huanjing de ruogan guiding* (The Regulations on Further Improving Soft Investment Environment for Foreign Investors), December.

The City Government of Meizhou (1998), *Meizhou shi guli waishang touzi de youhui zhengce he cuoshi* (Policies on Favourable Conditions for Foreign Investors in Meizhou), January.

The Commission for Foreign Economic Relations and Trade, Chaozhou (1998), *Yazhou jinrong weiji dui wo shi waijingmao ji dongnaya qiaozi qiye de qingkuang huibao* (A Report of the Impact of the Asian Crisis on the Foreign Trade and Southeast Asian Chinese Invested Enterprises), May.

The Commission for Foreign Economic Relations and Trade, Jieyang (1998), *Yazhou jinrong weiji dui wo shi dui waijingmao de yingxiang ji duice* (The Impact of the Asian Financial Crisis on the Foreign Trade and Economy in Jieyang, and the Policies of How to Deal with It), April.

The Commission for Foreign Economic Relations and Trade, Longhu (Shantou) (1998), *Yazhou jinrong fengbao dui wo qi jingji gongzuo de yingxiang jiqi duice* (The

Impact of the Asian Financial Crisis on the Foreign Trade and Economy in Longhu, and the Policies of How to Deal with It), April.

The Commission for Foreign Economic Relations and Trade, Meixian (1998), *Meixian waizi qiye qingkuang huibao* (A Report on the Current Situations of Foreign Invested Enterprises in Meixian), May.

The Commission for Foreign Economic Relations and Trade, Meizhou (1997), *Dongnaya jinrong fengbao dui wo shi waijingmao gongzuo yingxiang de qingkuang hubao* (A Report on the Impact of the Southeast Asian Financial Crisis on the Foreign Trade and Economy in Meizhou, and Policies of How to Deal with it), December.

The Commission for Foreign Economic Relations and Trade, Punning (1998), *Guanyu yazhou jinrong fengbao dui wo shi dongnaya qiaozi qiye yingxiang qingkuang de huibao cailiao* (A Report on the Impact of the Asian Financial Crisis on Southeast Asian Chinese Invested Enterprises in Punning), May.

The Commission for Foreign Economic Relations and Trade, Shanwei (1998), *Shanwei shi qiaozi qiye shou yazhou jinrong fengbao yingxiang jianyao qingkuang huibao* (A Brief Report on How the Overseas Chinese Invested Enterprises were Affected by the Asian Financial Crisis in Shanwei), April.

The Guangdong Provincial Commission for Foreign Economic Relations and Trade (1998), *Zhenfen jingshen, kaituo qianjin, quebao wancheng jinnian de weijingmao renwu* (To Lift Your Spirits Up and Move Forward to Ensure the Fulfilment of the Goals of Guangdong's Foreign Trade and Economy in 1998), May.

The Guangdong Provincial Liaison Committee for Compatriots from Overseas, Hong Kong, Macao, and Taiwan (1998a), *Dongnaya jinrong weiji dui wo sheng qiaozi qiye yingxiang de diaocha baogao* (An Investigation Report on the Impact of Southeast Asian Financial Crisis on Overseas Chinese Invested Enterprises in the Province), May.

The Guangdong Provincial Liaison Committee for Compatriots from Overseas, Hong Kong, Macao, and Taiwan (1998b), *Guanyu zuazhu jiyu, jinyibu tigao liyong waizi he waimao shuiping de jianyi* (Suggestions on How to Grasp the Opportunities, and Further Increase the Foreign Investments and Foreign Trade), May.

The Guangdong Provincial Office of the Overseas Chinese Affairs (1998a), *Guanyu dongnaya jinrong fengbao dui chaoshan diqi san chi zaocheng yingxiang de diaocha baogao* (An Investigation Report of the Impact of the Asian Financial Crisis on Overseas Chinese Invested Enterprises), May.

The Guangdong Provincial Office of the Overseas Chinese Affairs (1998b), *Taiguo zaosou jinrong fengbao chongji houguo yanzhong* (The Serious Consequences of the Financial Crisis in Thailand), April.

The Guangzhou Daily.

Huang, C. (1998), 'Overseas Chinese and China's Economic Modernisation' in Paul van der Velde and Alex McKay (eds), *New Developments in Asian Studies*, Kegan Paul International, London, pp.123-139.

The Longhu District (Shantou) Office of the Liaison Committee for Compatriots from Overseas, Hong Kong, Macao, and Taiwan (1998), *Guanyu dongnaya jinrong fengbao boji muqian waijingmao zhengce, jinrong zhengce tiaozheng hou dui wo qi waijingmao gongzuo de yingxiang ji yingdui banfa* (The Impact of Southeast Asian Financial Crisis on the Current Policies of Foreign Trade and Finance in the District), May.

The South China News.
Wang, X. (1998), 'Exports Get Boost from Tax', *South China Morning Post*, 12 June.

8 Responses to the Crisis: Network Production, Diversification and Transnationalism

DAVID IP

Introduction

When the 'Asian Crisis' first struck Thailand in July 1997, then spread rapidly to the neighbouring countries of Indonesia, Malaysia, Singapore, then to South Korea, Hong Kong and Taiwan, the prevailing view was unanimously pessimistic and cynical. Most 'crisis-watchers' believed that the 'economic miracles'these Asian countries had achieved through the previous boom years were in fact bubbles. Their economic outlook and social climate were at best doubtful and uncertain. Today, almost three years down the track, although most 'crisis-watchers' have become more cautiously upbeat, a sustained recovery is still considered less than certain.

Chanda's report (1999/2000, pp. 20-23), for example, was full of good news. He reported that in late November last year the Asian Development Bank had forecast that the Asian region would have a 5.7% economic growth rate at the end of the year instead of the sluggish 2.3% in the previous year. Simultaneously, the Bank also saw that because of falling prices for consumer goods and in spite of widespread wage cuts, real incomes might not have eroded as much as had been feared. Moreover, South Korea is projected to have a growth of 9%, followed by China at 7.2%, and laggardly Hong Kong, was finally looking to a 1.8% growth. Yet Goad insisted that 'for all the surprising good economic news of recent months, and there has been plenty, Asia's recovery is still spotty' (1999/2000, p. 106). In particular, he charged that, except for a few smart companies, most businesses and governments in Asia were blissfully unaware that they would have a tough time in the future competing with other parts of the world that were becoming more liberalised and competitive – for example, with Poland setting the pace in Eastern Europe with growth rates in the 5-6% range, and in Latin America where Brazil had attracted US$27.11 billion of net FDI in the first 11 months of 1999 alone, up sharply from US$23.45 billion in the same year period in 1998. Furthermore, he felt that the export of the Nasdaq to Asia and the recent rushes to embrace the Internet, while an indication of the survival of Asia's entrepreneurial spirit, could turn into another financial indigestion or stock market bubble.

Similarly, although some larger Asian corporations had rebounded and recovered in 1999, as indicated by the most recent annual listing of the 500 largest corporation in Asia by *Yazhou Zhoukan,* the journal also pointed out that the total number of enterprises that suffered losses in 1999 had also increased three fold from 30 in the previous year to 93 in the current year. Moreover, in spite of the fact that the total market value of the top ten largest corporations had risen 80% from the previous year, the last 10 enterprises on the list could only manage to increase their value by around 3%, and were still 30 times below what they were worth before the Asian crisis struck. This, the journal thought, indicated clearly that the smaller [of the largest] enterprises had limited capabilities to respond effectively to the crisis. In these contexts, it is timely for us to examine the business strategies diaspora Chinese have adopted during the times of the Asian crisis.

In this chapter, our objective is to sketch a preliminary picture of how Chinese entrepreneurs in Hong Kong, Taiwan and Australia have responded to the Asian crisis based on our research in progress[1]. We are specifically interested in finding out how the crisis has impacted on the diaspora Chinese businesses. For example, have they lost their autonomy and been forced by the circumstances to abandon their firm grip on the union of ownership, management and control of their enterprises? In responding to the crisis, have they sought assistance from their networks, long established or otherwise, as a particular crisis strategy for business survival? In production, have they retreated from diversification and retracted into their core business? As a strategy, has transnationalism been a source of unsustainable risk or a source for recovery?

The information we are presenting here was gathered from in-depth interviews with 25 informants in Hong Kong (12), Taiwan (5) and Brisbane, Australia (8). In Hong kong we attempted to re-interview two of the informants who had participated in our earlier study conducted between 1994 and 1995. Both had suffered heavy losses during the Asian crisis. Only one, however, gave his consent to be re-interviewed. Of the eight informants we interviewed in Brisbane (mostly immigrants originating in

Hong Kong and Taiwan with one from mainland China), half (4) were previous participants and all had been doing well in their businesses. Of the remaining four, two had expanded while the other two had setbacks. The informants in Taiwan were all new participants.

Impact of the Asian Crisis

From any major crisis, there are winners as well as losers. While many firms have failed, there have been far more survivors. This is perhaps the most appropriate way to summarise how the Asian crisis has impacted on our group of entrepreneurs.

Most striking was that the majority of our informants (most of our informants were manufacturers while only a few were involved in trade or services) claimed they had not been affected severely by the crisis at all. Only three, two in Hong Kong and one in Brisbane suffered heavily. The closure of two businesses, by the admission of their owners, was not, however, directly the result of the Asian crisis. A garment manufacturer in Hong Kong admitted that

> The closing of my business has been a slow death. It started even before the Asian crisis hit the region. It had been a sunset industry with decreasing profit margin over the years and not even relocating my operation to mainland China helped. One day when this person from the same township where I operated my garment factory walked into my office with a suitcase full of shirts, I knew right away I had to fold up my business. He spread all the shirts in front of me on my desk and said to me bluntly, 'I could make these shirts for 8 *yuan* for you'. I couldn't even get them made for less than 10. How could I compete with him? Did I manage my business badly? I have no way in knowing because I tried my best, but I was certain that I no longer could compete with mainlanders who had learned every trick from their Hong Kong 'masters'and beat them hands down.

A retailer who closed her high fashion boutique in Brisbane similarly blamed her own misjudgment rather than the Asian crisis.

> I had planned to start my boutique shortly before the Asian crisis as a way of diversifying my regular business [exporting live prawns exclusively to the Japanese market] which I have been running since my arrival in Australia about 8 years ago. When the crisis started my friends advised me not to go ahead with it but I thought it would not be hitting Australia

and I was not going to limit my clientele solely to the Hong Kongers or Taiwanese. I thought there was a niche market for mainstream Australians as well so I went ahead as planned. During the first six to eight months the business went really well but after that, it just went dead. No Chinese or Australian customers walked into the shop. In hindsight, I suppose no one was in the proper mood to splurge on high fashion when there was a crisis; and things began to look a bit uncertain. I had to admit I had misjudged the situation.

These two were small and medium sized businesses. The third enterprise that folded, ironically, was a much larger corporation that had prided itself on its rapid rise from a small family business to a transnational conglomerate that manufactured industrial ceramics. The owner of the business declined our request for a re-interview. However, how it went bankrupt in only a few years time was well reported by other informants we contacted, amongst tales of 'unsustainable rapid expansions in property speculation, unrestrained borrowings at the expense of liquidity and bad management'. Many regarded this case typically as an example of 'inflated ambitions', but some of our Hong Kong informants also acknowledged that his demise was aided by banks that should be blamed for having a similar interest in making quick, short-term profits.

In fact, informants were quite critical not only of the banks' impersonal approach and sudden decision to tighten credit to manufacturers after the Asian crisis struck, but also of the lack of interest among the banking industry in supporting the manufacturing sector. For many manufacturers, the term 'manufacturing' (*shi ye*), means 'real business', something in which they take enormous pride and regard as the fundamental blocs in an economy. Yet instead of support, they were given a squeeze on credit.

> It is ironic that the banks decided to penalise us by lowering our credit limits immediately after the Asian crisis spread to Hong Kong. We have always been the good guys who run our business honestly and diligently with little intention of borrowing money to speculate on properties or the stock market. Of course the values of the collaterals [our properties] the banks are holding have dropped because of the collapse of the property market, but we are not the bad guys who incurred bad debts. Now they are telling us that our credit will be limited at the very time when we most need the extra capital to take advantage of the reduced costs of production brought by the Asian crisis to expand our activities. They never seem to have any empathy for the difficulties they have brought us. Had we not build up our own reserve, we would be in deep trouble.

The recent move by the Hong Kong government to set up a credit scheme to assist small and medium sized enterprises to ease the pain imposed by such credit squeeze was welcomed by most informants, although some decided that the plan was too limited, too late and under-publicised. An informant who set up her import-export business immediately after the crisis struck, for example, complained about the lack of enthusiasm from bank managers when she approached them for details about the scheme.

> The banks simply weren't interested in this because they couldn't make much money from this and they tried to brush me off. They simply said they knew nothing about it.

Others were equally irked by the lengthy period of time they had to wait to obtain credit from the scheme and decided that 'self-reliance' was the best bet to ride out the current financial storms.

Autonomy, Management and Control

The great majority of the entrepreneurs we have interviewed so far had managed to escape severe damage from the Asian crisis, especially in terms of their autonomy and control of their businesses. For many of them, however, the significance of the Asian crisis was that ultimately it screened out those businesses that had been operating shakily for quite a while. They believed the Asian crisis had provided an easy excuse for those who failed and who were not willing to reflect on how badly they had managed their business. Nevertheless, the crisis and its aftermath have also prompted them to think seriously about whether it would be necessary to have professional managers and about the urgency of technology upgrade to better their chances of survival in the increasing competitive world of globalisation and rapid changes.

In Qin's (1998) book on the 21 'pioneering' entrepreneurs in Taiwan's electronics industry, over half (13) recognised the important role of professional managers in the management of their businesses. Among our informants, however, such a view is not prevalent. Those featured in Qin's interviews were larger corporations, while most of the businesses operated by our informants were, in their own words, 'small operations' and thus, 'there was no room for a professional manager to play a significant role'. Or, as one informant put it rather bluntly, 'we can't afford to put another person on our payroll. We're more concerned about the bottom line of staying in business'. On the other hand, however, many of our informants also seemed to agree that 'when a company becomes publicly listed, in order to maintain its reputation and credibility, professional managers would be appropriate'.

Nevertheless, a more sensitive question remains – had their business grown bigger and it became necessary for them to have professional managers on deck, how would they handle the issue of control of their businesses while maximising autonomy vis-à-vis outsiders? Here we detect a sense of uneasiness as well as doubt about the professionalisation of management. A manufacturer who had revolutionalised the use of electroluminescent boards in advertising and promotion, for example, was less than complimentary about his managers.

> I have managers in my office but they are not as important as the members on my creative team. They are the ones to translate my ideas into reality, but not the managers. Perhaps because I run a small business, I find it easier for me to maintain major control like dealing directly with my clients, overseeing the update of our web page on the Internet, and planning future directions and strategies. I'm the real professional here because I know my business inside out, from its technology to its market and production. Where do you find a manager who knows all these?

Hiscock's (1999)[2] prediction – that in the age of the current climate of mega-mergers, family-owned enterprises perhaps could end up in a lonely death if they didn't find a partner – also did not seem to be an urgent matter for our respondents.

Social Networks

Most of our informants did not suffer a great deal from the fallout from the Asian crisis and few had had to seek financial assistance from their social networks. Did they know anyone from their own networks who was less fortunate and was hard hit by what came with the crisis? Only a few reported they had 'heard stories' about isolated cases of misfortune of friends or contemporaries who needed financial support or rescue, while none reported incidents of people actually coming forward requesting them for help.

Does this mean that social networks work better in good times than bad? Or, are networks more instrumental and opportunistic in nature?

Observers have tended to implicate the use of networks in Asia as part and parcel of 'crony capitalism' (Backman 1999), but judging from the information from our respondents, the use of networks in their business context is in fact perceived as highly moral with its own implicit rules about reciprocity, responsibility, business ethics and control.

> You don't abuse the trust and people in your network because the words will spread quicker than you know if you do. You'll lose not only 'face', but ultimately also your credibility as everyone you know in the network finds out you've done something wrong. For example, if you have difficulties in making payments to your supplier within the credit period, you'd ask for an extension rather than running away. You run away once and you'll never be trusted again. On the other hand, if you're honest, as we all run small businesses, we understand the difficulty people have particularly during times of crisis. That's why I don't even wait for people I know in my network to approach me asking for extension of time to make payments. I just told them to take an extra month or two. Why make things difficult for them if in the end they come back to you with troubles? (Hong Kong audio-tape manufacturer).

> Before the crisis, I knew a number of people who made a lot of money from speculating properties and on the stock market. Some rang and asked me to invest 'smartly' with them. But I declined because I was angry with them. I told them, 'you're a manufacturer and you know what you're doing is going to make things more difficult for us? Who doesn't like quick, handsome profit? But what you've done will make rent more prohibitive and production costs skyrocketing. Then no one wants to work honestly any more in manufacturing. 'Of course, there were times I found it real tempting and easy to follow what they did. I'm only too glad that I didn't because I would have put my own business at risk. I now hear that they are in financial trouble, but they would not dare to ask anyone in the network to help them out. Why? Because they ought to be ashamed of what they did (Hong Kong manufacturer of soft luggage and bags).

> Few people would ask you for monetary help during this time because they also know that things are tight all round. Mostly the support they needed was an extension of credit – more time to pay for the spare parts or raw materials they need. I think that's fine because it's really about business. You don't want to help them or anyone who come to you to borrow money to pay their gambling debts (Taiwan manufacturer of chemical dye).

Production Activities and Diversification

The significance of social networks in times of adversity is thus apparent. It obliges mutual support but mainly for keeping businesses or production going. It has a moral code that discourages extending support to members who are in trouble because of personal misdemeanour but is readily supportive of those who stick by the business ethics. To a number of our entrepreneur informants, 'running a business is not just about making your own profit. It is also about keeping the industry going'.

The Taiwanese entrepreneurs had learned to develop their social networks one step further into a system of 'integrated production networks' – running compatible enterprises in a local area as a group, each specialising in the production of parts for a manufactured product like a personal computer or a mobile phone. In having their businesses located near one another, they formed an integrated network that supported one another while simultaneously cutting down transportation and production costs and saving valuable time in meeting export deadlines. Furthermore, as the commodity chain matured, the manufacturing of the components or spare parts had also become more specialised, thus allowing them to move up technologically to upgrade their production by concentrating on producing one or a small range of components of the production process. This further specialisation enabled them to save money, as less credit was required to finance their production and they became more competitive on the international market when their costs were lowered.

The transplant of this system into mainland China has enabled the Taiwanese entrepreneurs to reap enormous benefits even at a time when the general economic climate in the Asian region was depressed (Gao et al 1997; see also next chapter in this volume). Their successes have also prompted two important questions:

- Does this mean that entrepreneurs from Hong Kong, who have been operating their businesses more in a guerilla fashion (Smart 1993), without the benefit of an integrated production network, have done less well during the crisis?
- Why isn't there a similar integrated production network set up among the Hong Kong entrepreneurs?

In terms of volume of business, most informants we talked to were surprised by how little they had been bruised by the crisis. The overall impression our informants had was that production was in all cases up rather than down, although a few also reported that some lines of production had been scrapped, not because they were losing money, but as both a timely and convenient measure to streamline their operations. Others indicated their cautiousness in taking steps to expand production off shore because of the uncertainly imposed by the crisis. Most Hong Kong informants recounted a short six-month period when things were confused and the fallout of the crisis was unknown. They however eventually gained much at the expense of other countries that had experienced some severe misfortune during the crisis.

> We thought we would have a tough time competing with those countries that devaluated massively their currencies. A couple of months later, however, we found that the credit squeeze in those countries was so bad that virtually no manufacturer could get any loan from any bank in their country to continue with their production. Suddenly we had a windfall of orders from their clients who told us they had to switch to us (Hong Kong manufacturer).

What was expected to be a crisis had turned out to be a boon for these Hong Kong entrepreneurs, although not without some setbacks. As the crisis spread to Hong Kong, international buyers also realised that they were in a strong position to dictate their terms. They wanted lower prices and better quality and the Hong Kong enterprises had to oblige, not too unhappily. This was because labour costs had gone down, the problem of labour shortage due to rapid and high turnover had also suddenly disappeared, and despite the banks' credit squeeze, most manufacturers managed to get extra credit from their network associates. There have also been a few surprises for other manufacturers in 'sunset' industries such as production of video and audio cassette-tapes and clock radios. The crisis has given these industries a new lease of life, at least for a few more years.

For example, it is hard to imagine what future there would be for video and audio cassette-tapes in the era of Internet, MP3 and DVDs. Yet the crisis has unexpectedly revived the demands for the more humble and affordable tapes, because suddenly many countries in Asia found themselves unable to 'catch up with the Jones in the Western world especially when they have lost their shirts'. According to this informant, even Samsung in South Korea had to rely on its Hong Kong subcontractor to produce video cassette tapes, as their own tape production facilities had closed down during the more prosperous times.

In stark contrast to the Taiwanese entrepreneurs who have been pulling all stops out to upgrade their operation, not only from OEM to

ODM, but also shifting into OBM and even JDM (Qin, 1998; CommonWeath Editors, 1999) some of our Hong Kong informants' fatalistm almost seems ironic. Similarly, when the Taiwanese are expanding their operations in China with their integrated production networks, it seems curious that many of the Hong Kong business people we talked to still preferred to go alone into China rather than as a collective. Some respondents had their own justification:

> The Taiwanese are very different from us [Hong Kongers]. It is necessary for them to operate in mainland China as a collective because they do not have our advantage – the convenience of physical proximity to China. But many of their business are dealing with complex components in high technology, they need to stick together to operate smoothly. For us, not many are involved in high tech production. Why is it so? They [the Taiwanese] had a lot of help from their government, not like us in Hong Kong. We've always been treated like an orphan. No one cares if your business is dead or alive. We always have to fend for ourselves. The government has not provided anything for us – no science park, no supporting institutions for technological development. I developed this special technique to slice up audio-tapes from big rolls to produce cassettes in the 1970s. I am conscious that the technology is now old, and with the rise of DVDs and digital camera, my business is doomed. The current windfall will probably give me an extra five to seven years but after that, what could I do? I often thought about changing directions, getting into a new production line, like manufacturing toys. But that's not easy because it costs too much to invest. Besides, it requires an entirely different set of network to operate efficiently and effectively in a new line production you know very little about. It would be easier to stomach all the dramas in switching production when you are young. It is just too late for me to change now after being in this business for 30 years (Hong Kong audio-tape manufacturer).

He was not alone to be caught in this dilemma, but he was fortunate that he had diversified into producing plastic cases for video/audio tapes. When compact disks began to replace vinyl records, he found himself sitting on top of a small gold mine -- producing plastic cases for various types of compact disks that have been gaining increasingly widespread popularity. Less optimistic was the clock radio manufacturer. He saw the time of his business coming soon to an end, although at present, the sales volume had not been diminishing drastically 'thanks to the continuing demands from other developing countries'. He had pondered the possibility of diversifying into another line of production, like toys or 3-in-1 clock radios with CD play function. He too had found the initial financial outlay prohibitive and more significantly, the cultivation of new personal cum production networks and acquiring the inside knowledge of a totally different industry 'both too late and too demanding'.

It seems easy for us to conclude that during times of crisis, most entrepreneurs would choose to be more conservative, preferring to devote concerted efforts to consolidate their core business rather than venturing into unfamiliar and uncertain territories of diversification. However, the conventional wisdom of not putting all eggs in one basket and the necessity to diversify one's business when the opportunity arises did not seem to be altered much by the current crisis. Although not every one of our informant had the same enthusiasm to diversify their business, and indeed some were more cautious than others, a sizable number of our informants were optimistic about their own efforts in diversification.

The owner-producer of electroluminescent boards in Hong Kong, for example, began marketing a 'technology transfer' program on franchise basis on the Internet to clients internationally because he saw that 'it was a necessity to keep one step ahead of others'. Similarly, a young computer business owner in Australia started another business selling bacteria-based, organic fertilizer to countries in Europe and Southeast Asia when he realised that 'the computer business in Australia had become too competitive and the profit margin kept diminishing'. The beef-exporter whom we interviewed four years ago in Australia, was now spending more time in Taipei than Brisbane, running two new businesses where he had recognised an empty niche in the market. These included a chain of cafés specialising in Cappuccino and muffins for the trendy set, and a business manufacturing LED lighting. Likewise, another participant, owner of a news agency, in our previous study in Brisbane, has two other successful sidelines -- a printing business specialising in Chinese publications, and a Chinese restaurant catering mainly for Australian clients in the gentrified suburb of New Farm. He is now embarking on his third business involving food processing in China.

Transnationalism

The capacity, even of small firms, to operate transnationally, through family and network members in different countries, is one of the defining features of any diasporic capital. Under the shadow of the current Asian crisis, many of the diaspora capitalists are showing no sign of retreating from regional or global perspectives. One of the four largest land developers in Hong Kong, New World Development, for example, despite its set back during the first 12 months after the crisis spread to the ex-colony, reinvented its business activities in 1999 and expanded into the Internet, telecommunications and digital technology. Its field of operations had also been determinedly transformed from local into regional as exemplified by its numerous joint projects in Chinese pharmaceuticals in Sichuan and Yunan Provinces (*Yazhou Zhoukan,* November 1-7, 1999 p. 27). The much reported ambitions of Richard Li, son of Hong Kong's number one magnate Li Ka-shing, to be Asia's biggest Internet tycoon, is another typical case (e.g. Spaeth, 2000, p. 32). The determination of Taiwan's Acer computers to edge out China's biggest computer manufacturer, Legend, is also indicative of the intensification of its global push (Wu, 1999). In Australia, Telstra's attempt to take over Ozemail, the largest Internet server in the nation, was foiled by the relatively unknown Eisa. Eisa is two-third owned by Wang, who only arrived from mainland China less than 10 years ago, and is illustrating again the undampened transnational ambitions of diaspora Chinese capitalists (Sydney Morning Herald, 2000)[3].

By comparison, most of our informants' businesses are smaller than those whose doings are recorded in the media. Nevertheless they seemed equally eager to expand regionally or globally. This was also true for those operating from Hong Kong. The electroluminescent board manufacturer, for example, had been successful in penetrating the US market and at present was aiming to expand his clientele through the Internet. A producer of soft luggage had been rewarded with more OEM contracts for brand-name designers from both the US and European markets. A toy manufacturer also found prosperity in the US market, producing toys for fast food chains and movie tie-ins. An import-export trader, who set up her own business just after the Asian crisis struck, had similarly reoriented her efforts to Europe and the US from southeast Asia and South America.

Likewise, our Australian informants were equally global in their outlook. For example, the failed boutique owner was in Shanghai setting up a restaurant business specialising in Australian seafood. As well, she planned to expand her export of fresh prawns to South Korea. The computer business operator, in addition to selling organic fertilisers to Europe, at the time of our interview, had also set up a distribution network in China not only for selling fertiliser, but also materials used for sound insulation and plumbing. Two of the five informants we talked to in Taiwan indeed remained critical of transnational prospects. They argued that the chances for small and medium-sized enterprises making it successfully in the regional or global market were less than optimistic.

> I am very critical about all these theories about going global or regional. Everyone who has anything to do with running a business in Taiwan thinks it is the magic formula for success and that's why everyone is rushing to expand into mainland China, Vietnam, Indonesia, Malaysia and the Philippines. The fact is, in the last 20 years, most of the so-called successful small and medium size businesses in Taiwan have not been managed professionally like many large corporations. They made some money and their egos got inflated. When they found others expanding overseas, they followed blindly thinking that they could not be left behind. Yet many know nothing about the cultures or the business practices in the foreign land and soon found themselves trapped in a bottomless pit that keeps draining their capital. Large corporations have better chances – they have better capital supply, and they have the manpower. So what if they lose money for one or two years, or have managers who failed and returned home after three or five years? They have already built a foundation there. Send another manager in and given time, they will succeed. Like President Enterprise in China. We small and medium size enterprises will find it tough to last that long.

Ironically, however, both had in fact expanded their businesses (chemicals, dyes and plastics) to Vietnam and China five years ago and they had survived.

Conclusion

At the beginning of the Asian crisis, it was popular among doomsayers to conclude that the Asian economic 'miracles' had vaporised into 'mirages'. As the economic gloss continued to fade, many further questioned not only the viability of the economic, but also of the social structure of the countries affected by the crisis. When the dust began to settle, Sheridan (1999, p. 10) declared that 'a certain amount of the miracle has been undone by the crisis, but not all'. Similarly Dicken and Yeung (1999, p. 119) have concluded that the Asian crisis has not put ethnic Chinese business firms under siege. Instead, windows of opportunities were opened for those who were able to strengthen further their business and networks locally as well as regionally. The picture we have garnered from our research in progress is not dissimilar.

In every crisis there are losers and winners. Stories of business closures and indebtedness were commonly circulating among our informants. Many typically considered those who failed as 'gamblers', or 'the overly ambitious', who had strayed from their core activities taking extreme high risks to speculate in the property or the share market. Yet as our research unfolded, it became clear that business failures were never as simplistic as what had been rumoured. For example, our boutique owner failed not because she had taken uncalculated risks. Rather, she had misjudged the magnitude of the delayed impact of the crisis on the higher end fashion market in Australia. The garment manufacturer from Hong Kong did not gamble away his fortunes by speculating in properties or shares. He folded his operation in mainland China because of his inability to cut down production costs and remain competitive against the local Chinese manufacturers.

We would like to think that the 'winners' in our research, those who have successfully weathered the financial storms, must have done something right in responding to the sudden disaster, especially given that the credit squeeze had been sudden and severe. Indeed, a range of strategies was engaged in. For example, extension of debt repayment time by manufacturers was granted to members in their personalised networks, out of grace and consideration not just necessity, to increase cash liquidity and business sustainability. Production activities were streamlined and expansion plans were put on hold. The formation of an integrated production network also helped to increase specialisation that in turn led to the minimisation of competition and up-front capitalisation. Nevertheless, they also had unexpected good fortune - some benefited from a windfall of contracts flowing from producers in countries which had suffered so severely and deeply from the crisis that they could not carry on with production. Others who were engaged in sunset industries such as audio/visual tape cassettes found a new lease of life.

Notwithstanding, this is not to say that the 'winners' were having it easy. For example, many were made aware by the crisis of the dilemma their business would have to confront - that in the new world of regional or global interdependence, the future of their family-based operation might have to incorporate a variety of outsiders that may include professional managers, or strategic allies to survive, perhaps at the expense of their autonomy and control. Similarly, while most continued to embrace the well-accepted philosophy of responding to spontaneous opportunities in regional and global markets, they were also keenly aware of the obstacles this entailed for smaller businesses. In seeking to expand through diversification, they could face particular difficulties in raising capital, in finding appropriate network connections and in getting access to necessary knowledge about production activities or foreign markets.

In sum, what has emerged from our preliminary findings is that many of the distinctive characteristics defining diaspora Chinese business (Tracy et al 1999) have continued to be relevant during the Asian crisis. Despite the differential impact of the crisis on firms of different sizes, kinds of management and activities, many respondents still preferred to maintain full autonomy and control. Networks have remained for them both morally honored and significant as a mechanism for minimising debts and vulnerability. Few indeed seemed to abuse the trust and goodwill associated with such networks. Moreover, diversification and transnationalsm have not been sources of weakness and unsustainable risks. Rather, in the current climate of gradual, albeit slow recovery, many still followed such strategies, while exercising more cautious and rational productivism than before and avoiding speculative and risky operations.

We do not pretend these impressions to be precise, accurate or final. In fact, they are preliminary and as yet untested, as they represent what has emerged only from a limited set of interviews and observations. The discovery that there are booms in doom amongst diaspora Chinese businesses however is at least indicative of the danger and fallacy of simplistic doomspeaks.

Notes

[1] The research, 'Diaspora Chinese Capitalism Responses to the Asian Crisis: A Longitudinal Study', is funded by an Australian Research Council (ARC) Large Grant for the period 1999-2001. It will include as many re-interviews as possible with the original entrepreneurs in Australia and Hong Kong who participated in our earlier study on diaspora Chinese capitalism, plus new interviews with business owners in these countries and in Taiwan and Singapore, selected by means of snowball sampling – an eventual total of 120 interviews.

[2] *The Australian*, 20 April 1999, pp. 32.

References

Backman, M. (1999), *Asian Eclipse: Exposing the Dark Side of Business in Asia*, John Wiley & Sons (Asia) Pte Ltd, Singapore.

Chanda, N. (1999/2000), 'The Year of the Internet', *Far Eastern Economic Review*, December 30 and January 6.

Dicken, P. and Yeung, H.W.C. (1999), 'Investing in the Future: East and Southeast Asian Firms in the Global Economy', in K. Olds, P. Dicken, P.F. Kelly, L. Kong and H.W.C. Yeung (eds), *Globalisation and the Aisa-Pacific: Contested Territories*, Routledge/Warwick Studies in Globalisation Series, London and New York, pp. 107-128.

Gao, X. (Kao, H.C.) et al (1997), *Taisheng Jingyan -- Touzi Dalu De Xianchang Baodao (The Taiwan Investment Experience in Mainland China -- A First-hand Report)*, Commonwealth Publishing Co. Ltd., Taipei.

Goad, G.P. (1999/2000), 'Catalysts for Change', *Far Eastern Economic Review*, December 30 and January 6.

Qin, M. (1998), *Kaichuang Taiwan Dianzichuan Renwu Fengyun* (The Pioneers of Taiwan's Electronics Industry), Daotian Publishing Co., Taipei.

Sheridan, G. (1999), *Asian Values Western Dreams: Understanding the New Asia*, Allen & Unwin, St. Leonards, NSW.

Smart, A. (1993), 'Gifts, Bribes and *Guanxi*: A Reconsideration of Bourdieu's Social Capital', *Cultural Anthropology*, 8(3): 388-408.

Spaeth, A. (2000), 'Dial M for Merger', *Time*, February 28, p.50.

Sydney Morning Herald, February 2000.

Tracy, N., Ip, D. and Lever-Tracy, C. (1999), 'Winners and Losers -- the Crisis Strategies of Diaspora Chinese Capitalism', paper presented at the Second Bamboo Network Seminar: Challenges to the Chinese Overseas in an Era of Financial Vulnerability in the Asian Pacific Region held at the University of New South Wales, Sydney, Australia, May 14-15.

Wu, Y.Y. (1999), 'Hongji Qianggong Dalu Tonglu' (Acer Attacks the Channels in Mainland China), *CommonWealth*, November, pp. 184-188.

Yazhou Zhoukan The Chinese International Weekly, November 1-7, 1999.

9 Weathering the Storm: Structural Changes in the Chinese Diaspora Economy

NOEL TRACY

The final decades of the 20th Century saw momentous changes in the East Asian region. Rapid and sustained export-oriented economic growth completely reshaped the landscape of the international economy making the region the principal engine of growth for both world GDP and international trade. In these decades the Chinese diaspora business communities throughout the region became a significant current in global capitalism, knitting much of the southern part of the region together and serving as a mentor and bridge for the entry of China into the international economy[1].

The financial crash that had engulfed the whole region from May 1997 onwards, and only began to show signs of coming to an end in the latter part of 1999, presented a major challenge for these developments. The crash revealed weaknesses within existing institutions, organisations and strategies, bringing many to the brink of insolvency and some to liquidation or seizure by governmental institutions. It also induced new strategies, produced fundamental changes of direction and revealed strengths. Although the full impact of the crisis has yet to be completely worked out, it is clear there have been winners and losers among the many Chinese business groups. Some large and well established players have gone under or seen their fortunes and prospects substantially reduced while others have prospered despite the difficulties. What is probably of greater significant in the longer term is the substantial number of new start ups that have occurred and the industries these are in. We can only speculate about how many of them will become the important corporations of the future but it is clear some have already made their mark. What has emerged from this watershed will shape an important part of the region's future.

Major economic crises produce major structural changes in the economies affected. The great Depression of the 1930s saw the establishment or consolidation of what were to be the key industries of the post-war world, such as automobiles, aircraft, petrochemicals and whitegoods (consumer durables) and the beginnings of the shift towards the new consumer-oriented economy. In the process, the United States, which had been the worst hit by the depression, was transformed and finally emerged as the world's strongest economy[2]. Similarly, the American crisis of the 1980s saw business leadership pass from the old Fordist Giants like General Motors, Ford and IBM to Microsoft, Intel, Sun Microsystems and Motorola. In the current crisis, it could be a rgued that it is the absence of these changes which is at the heart of the Japanese and South Korean crises. There has been no changing of the guard and the giants which have dominated these economies since the 1960s now suffer from considerable overcapacity in the industries in which they excel like automobiles and consumer electronics. In the process, Japan appears to have lost its way. The Japanese and South Korean problems have to be seen as long-term structural crises, meaning that long-term revival requires fundamental changes to both the structure of their economies and the societies which spawned them.

This book, however, is not about the crises in Japan and South Korea, interesting as they are, the questions we have to ask are about Chinese diaspora business groups. How badly have they been affected? which groups have gone under and which have emerged stronger? and which of the new companies might be the leaders in the next decade?

Information about Chinese diaspora business is not readily available. This is to some extent because of their opaque business practices but is most particularly because so much data collection takes place at the level of the nation state. The process of globalisation has gone a long way ahead of the practice of data collection. Economies may no longer correspond to national boundaries in practice but the practice of data collection at both the national and the international level continues to assume that they do. This makes it difficult to assess even the overall impact of the crisis on Chinese business, in those national economies wherein they constitute a significant part of the private sector, purely from the national data, in the same way as you might be able to proceed, albeit with caution, in the cases of Japan and South Korea.

We do, however, have two significant sources of data for the larger corporations. By far the most important is annual list of the 500 largest Chinese privately-owned business groups outside China produced by the international Chinese-language weekly *Yazhou Zhoukan* from Hong Kong. The editorial staff of this publication maintain a close monitoring of the largest groups, for which they are to be commended. This is far and away the most important data base on the activities of the Chinese business communities and is an invaluable source of information. The second source is the various lists of the world's and the region's richest business families produced by a number of business publications including *Fortune*, *Forbes* and *The Australian*. These latter lists are more useful than might appear at first glance. With the majority of Chinese business corporations remaining under the control of entreprenurial families and very few passing into institutional ownership and its ensuing separation of management and control, personal wealth can be identified with control of business empires far more widely than it can in North America, Europe or Japan. Despite the increasing number of Chinese corporations listed on the region's stock exchanges, the majority remain tightly under the control of entrepreneurial families[3].

An analysis based on the largest corporations and wealthiest business families does, however, have its limitations. The real strength of Chinese diaspora business as we have argued elsewhere is its capacity for regeneration (Lever-Tracy, Ip and Tracy, 1996). The period of fastest growth in the 1980s and 1990s was one in which the tendency was for expansion and growth in the southern half of the East Asian region to occur by the multiplication of productive units rather than by the consolidation of existing business into larger groups. Chinese business tends to create entrepreneurs while Japanese business tends to create large organisations: two entirely different 'ideal types' of business construction (Sit and Wong, 1989; Redding, 1990, 1991; Clegg and Redding, 1991; Greenhalgh,1988; Hamilton, 1991).

New entrepreneurs are constantly emerging to replace those who have either lost their dynamic or made mistakes that ended their business empires and aspirations. A problem, therefore, is that an analysis of the giants will tell us little about new companies, the process of regeneration in practice as ex-managers and employees set up on their own. Another difficulty is that an analysis of the largest corporations does not give us a picture of what is happening to business groups. This creates particular difficulty in analysing Chinese business. The preference for a number of separate companies rather than consolidation into a single business organisation is another hallmark of Chinese business. A business group with a large number of separate companies may well not contain a single one which would make the top 500 but at the same time the combined strength of the group might rank them in the top 100 regional business groups.

These reservations apart, the analysis contained in the *Yazhou Zhoukan* compilations does give us an indication of the quite profound changes that have taken place in what we call the Chinese business sphere, basically, Hong Kong, Taiwan and Southeast Asia[4]. The most obvious change is in the economic geography of the Chinese sphere. Taiwan has dramatically increased in importance. While a third of the largest corporations were in Taiwan in 1995, in 1999 it was almost half (48%). On the other side of the equation while 213 (43%) of the largest companies were in Southeast Asia in 1995, in 1999 there were only 155 (31%). The data for Southeast Asia, however, needs some further disaggregation because the impact was by no means evenly spread.

The number of these large companies in Singapore has increased substantially, from 44 to 67 while the number in the three large economies particularly badly hit by the crisis, Indonesia, Malaysia and Thailand has dropped dramatically, by more than half from 155 to 75. Another substantial change is the rise of Singapore as a centre of Chinese business. With its economy for so long dominated by multinational corporations and the large state sector, it was never in the past able to rival Hong Kong as a centre of Chinese business activity. A glance at table 9.1, however, will reveal quite an important change in relative importance. Prior to the crisis, Hong Kong had nearly three times as many of these large companies based there as Singapore (120 against 44), now it has barely 50% more (105 to 67).

What do these changes represent? First, Taiwan, essentially because of its position as the industrial core of the Chinese business sphere, survived the crisis far better than other parts of the region. Taiwan, despite being forced to devalue its currency early in the crisis and re-impose elements of capital controls, managed to maintain economic growth at 4-5% per annum throughout the crisis. Second, Singapore although initially hit very hard by the crisis because of its proximity and exposure to two of the most badly affected economies, Indonesia and Malaysia, was the first to recover. Hong Kong was in an intermediate position. It was badly affected by the crisis with its stock exchange, property and retail sectors particularly badly hit. The value of shares on its Stock Exchange had fallen by more than 60% at the height of the crisis. Hong Kong, however, survived without major bankruptcies and its economy recovered in the last quarter of 1999. Despite many forecasts of doom about Hong Kong's future, its economy showed considerable resilience throughout the crisis. This was particularly true for its financial sector, with no commercial bank having to declare insolvency or seek government support despite major exposures to badly affected economies and corporations.

This was far from the case in the most badly affected economies, Indonesia, Malaysia and Thailand. Here, Chinese-owned corporations fared badly, with banks, manufacturing and commercial enterprises going under. The most spectacular collapse was the Bank of Central Asia in Indonesia. Jointly-owned by Liem Sioe-leong and members of the Suharto family, it succumbed to bad loans and foreign borrowings and had to be taken over by IBRA, the entity set up by the Indonesian government to administer insolvent banks and financial institutions. Likewise, the seizure of the Bangkok Metropolitan Bank by the Bank of Thailand in January 1998, following its collapse, wiped out the fortune of one of Thailand's oldest and richest Chinese families, the Techapaibuls (*Far Eastern Economic Review*, 2 February, 1998).

The Wanglee family in Thailand possibly only escaped the same fate by selling Thai Nakorthon Bank to British banking giant Standard Chartered (Straits Times, 29 April 1999). It should be noted, however, that while Standard Chartered is British-based and managed, its single largest shareholder is Singapore tycoon, Khoo Teck-puat. The largest Thai conglomerate, Charoen Pokphand, with major interests in agribusiness and telecommunications, was forced to sell its supermarket chain, Lotus, to the UK's Tesco. In Malaysia, all Chinese-owned banks are to be forced into mergers following the crisis as part of the Malaysian government's plan to restructure the financial sector into no more than 10 large groups, although at least two are likely to emerge as lead banks in the new groupings. William Cheng's Lion Group, headquartered in Malaysia with regional interests in steel and energy, ran into severe difficulties resulting from its over-ambitious expansion in the 1990s.

The overall impact of the crisis on Southeast Asia's Chinese business corporations can be seen from table 9.1. Seventy seven of the largest Chinese-owned corporations were headquartered in Malaysia in 1995, in 1999 there were only 50, a fall of more than a third. In Thailand the position is even worse. There were 45 in 1995 and only 13 in 1999. In Indonesia, only 12 of the 36 in 1995 were left in 1999. In the Philippines, the situation was not so bad, reflecting the lesser impact of the crisis, with the numbers making the list little changed.

Despite the difficulties encountered in the crisis, the vast majority of the region's largest Chinese business families have survived, albeit with some of them substantially reduced in wealth and in both the size and scope of their business empires. Others, however, have substantially increased their business power. The recently published list of the 100 richest business families in Asia, contained no fewer than 58 Chinese diaspora business families, a quite remarkable and disproportionate number, given that the total number of ethnic Chinese in the East Asian region outside Mainland

Table 9.1 The 500 Largest Chinese-owned Private Companies 1995-1999

	1995	%	1997	%	1998	%	1999	%	Increase (decrease) 1995-99
Hong Kong	120	(24)	113	(23)	105	(21)	105	(21)	(15)
Taiwan	167	(33)	183	(37)	285	(57)	240	(48)	73
Southeast Asia	(213)	(43)	(204)	(41)	(110)	(22)	(155)	(31)	(58)
Singapore	44	(9)	49	(10)	37	(7)	67	(13)	23
Malaysia	77	(15)	91	(18)	41	(8)	50	(10)	(27)
Thailand	45	(9)	15	(3)	12	(2)	13	(3)	(32)
Philippines	14	(3)	15	(3)	11	(2)	13	(3)	(1)
Indonesia	33	(7)	34	(7)	9	(2)	12	(2)	(4)

Source: Yazhou Zhoukan 1995-1999.

China number little more than 50 million against 130 million in Japan, 44 million in South Korea and well over 400 million in Southeast Asia. Against the 58 Chinese families, there were 18 Japanese, 6 South Koreans, and a total of 11 from Southeast Asia. There were 7 Indians (The Australian, 27 January 1999: 18). The 58 Chinese families were distributed as follows: 15 in Hong Kong, 11 in Taiwan, 8 in Malaysia, 8 in the Philippines, 6 in Singapore, 6 in Indonesia and 4 in Thailand. All had personal interests in net assets in excess of US$1 billion in the corporations they controlled (see Table 9.3).

The fate of the fortunes of the largest families as a result of the crisis are exemplified by the two families generally seen as the richest in Hong Kong and in Southeast Asia, Li Ka-shing's in Hong Kong and Liem Sioe-leong's in Southeast Asia. Despite setbacks in the Hong Kong property sector, Li's empire has gone from strength to strength. With major interests in container ports world-wide and telecommunications, his group has been able to survive the crisis intact, take new initiatives and make new aquisitions. Even during the height of the crisis the market value of his flagship company, Hutchinson Whampoa, had increased by more than 50%. This is in stark contrast to the fate of Liem Sioe-leong's regional empire. The insolvency of many of his corporations in Indonesia and the collapse of his core bank, Bank Central Asia, have substantially reduced his business interests. A target of anti-Chinese anger in the riots that brought down Suharto, because of the two men's close links, he was forced to flee to the United States leaving his son Anthony to try and revive the family's Indonesian interests. Bank Central Asia was taken over by IBRA and Liem was forced to pledge his interests in more than 100 of his other companies to IBRA in return for being allowed to continue to operate his companies when they were unable to repay their debts to the now nationalized BCA and other state banks (*Far Eastern Economic Review*, 1 October 1998). If they are to recover control of these companies, the family will have to buy back the shares handed to IBRA at the height of the crisis.

It is, however, not all doom and gloom for the Liems. Their core business in Hong Kong has come to the rescue. Responsible for at least a third of the group's profits even prior to the crash, listed First Pacific Group has bought out a number of Liem's Indonesian companies, thus enabling the family to discharge debt, and in 1999 it began to engage in a series of acquisitions on behalf of the group. The most notable of these was the purchase of Philippines Long Distance Telephone, which, when added to its existing telecommunications interests, made First Pacific the largest operator of both fixed-line and mobile telephones in the Philippines. As we come out of the crisis, it is becoming increasingly clear that the survival of the Liem group depends very much on its Hong Kong company and the latter's regional interests, which stretch throughout Southeast Asia, into Australasia and as far as North America. First Pacific, a component stock of the Hang Seng Index, has also seen its share price soar since the beginning of 1999. In Indonesia itself, the Liem family is no longer even the richest Chinese business family there, let alone the richest in the whole of Southeast Asia.

The success of Li Ka-shing's group and the survival of the Liem group provide us with some clues as to how businesses have ridden out the crisis. As a hypothesis that still requires further investigation, it appears to be the case that those groups that have taken advantage of and seized the opportunities provided by deregulation, regionalisation and globalisation have survived better than those who had chosen to remain concentrated on or too dependent on the domestic economy of their home base.   There are a few clear exceptions to this, the Indonesian cigarette manufacturers, Gudam Garam, Djarum Kudus and HM Sampoerna, controlled by the Halim, Hartono and Putera families respectively, Chinese despite their Islamicized names (their Chinese family names are Tjoa, Hwie and Lim) have clearly prospered despite the severity of the crisis. The case, however,

Table 9.2 The Largest Chinese-owned Technology Companies 1997-1999

	Rank in Chinese 500			Base	Market Capitalisation 1999 (US$bn)
	1999	1998	1997		
Taiwanese:					
1 Quanta Computers	5	-	-	Taiwan	13.9
2 United Microelectronics	6	7	6	Taiwan	13.8
3 Ausastek Computers	7	5	96	Taiwan	12.9
4 Hon Hai Precision Indust	15	20	53	Taiwan	6.6
5 Acer Inc	16	21	20	Taiwan	6.6
6 Advanced Sanicond. Eng	20	16	51	Taiwan	6.0
7 Compal Electonics	24	25	112	Taiwan	4.6
8 CMC Magnetics Corp	26	72		Taiwan	4.5
9 Inventee	29	18	42	Taiwan	3.8
11 Winbond Electronis	36	28	55	Taiwan	3.3
12 Arima Computer Corp	50	-	-	Taiwan	2.5
13 Siliconware Pres. Ind	57	57	-	Taiwan	2.1
Other:					
10 Johnson Electric	30	41	135	Hong Kong	3.8
14 Telecom Asia	58	172	47	Thailand	2.1
15 Creative Technology	65	65	94	Singapore	2.0
16 Shin Corporation	98	206	157	Thailand	1.2
17 Delta Electronics	141	165	377	Thailand	0.8
18 V-Tech Holdings	153	102	369	Hong Kong	0.7
19 Shin Satellite	275	-	-	Thailand	0.3
20 Gul Technologies	340	-	-	Singapore	0.3

Note: Pacific Cyberworks & Pacific Internet with initial IPOs in 1999 will enter the list in February 2000.
Source: Yazhou Zhoukan.

generally holds true across the whole region. A very clear example of this is the Quek family, which controls the Malaysian branch of the Hong Leong group. Severely mauled by the severity of the crisis in Malaysia and hit again a year later by the political fallout from Dr Mahathir's sacking and subsequent persecution of Anwar Ibrahim, with whom the group's chairman, Quek Leng-chan, was closely associated, the group and the family fortune nonetheless came out of the crisis intact as a result of its regional spread. While its Malaysian bank, Hong Leong Bank was in danger of being forced into an unwanted merger, with a possible loss of control, as part of the government's plans to restructure the financial sector, the group's largest Hong Kong bank, Dao Heng (the other is the Overseas

Table 9.3 Chinese Billionaires in 2000

1.	Kwok Brothers	HK	Sun Hung Kai Prop	10.0
2.	Li Ka-shing	HK	Hutchison	10.0
3.	Tsai Wan-lin	Taiwan	Cathay Life	7.7
4.	Lee Shan-kee	HK	Henderson Land	6.0
5.	Wang Yung-ching	Taiwan	Formosa Plastics	4.9
6.	Cheng Yu-tung	HK	New World Develop.	4.3
7.	Wu Tung-chin	Taiwan	Shin Kong Insurance	4.1
8.	Robert Kuok	Malaysia/HK	Kerry Group	4.0
9.	Kwek Leng-beng	Sing/HK City	City Develop	3.3
10.	Lim Goh-tong	Malaysia	Genting	3.2
11.	Wang Nina	HK	Chinachem	3.0
12.	Ng Teng-fong	Sing/HK	Far Eastern Org	2.9
13.	Quek Leng-chan	Malay/Hk	Hong Leong	2.5
14.	Fok, Henry	HK/Macau	STDM	2.5
15.	Khoo Teck-puat	Singapore	Goodwood Hotel	2.4
16.	Rachman Halin			
	(Tjoa To-hing)	Indonesia	Guadong Garam	2.2
17.	Chearavanont	Thailand	Charoen Pokphand	2.0
18.	Tan, Vincent	Malaysia	Berjaya	2.0
19.	Hsu Yu-siang	Taiwan	Far Eastern Group	2.0
20.	Ho, Stanley	HK/Macau	Shun Tak, STDM	2.0

HK 7, Taiwan 4, Malaysia 4, Singapore 3, Indonesia 1, Thailand 1, Property 7, Finance 2, Conglomerates 5, Manufacturing 2, Gambling 4.

Source: The Australian, Asia's Billionaires in 2000, 27 January 2000.

Trust Bank) was in the course of being made a component stock of the prestigious 33 member Hang Seng Index in December 1999. This latter represented a substantial recognition of the group's increasing importance in the Hong Kong financial sector.

One of the assumptions of many analysts at the height of the crisis was that substantial numbers of Chinese entrepreneurs would be forced to sell their corporations or at least dispose of significant proportions of them to western interests. Hiscock put this succinctly when he wrote 'Asia's family-owned conglomerates could face a lonely death if they dont find a partner' in the new world of mega-mergers (*The Australian*, 20 April 1999). It was assumed that the crisis was so severe that the possibility of recapitalizing their businesses, let alone remaining internationally competitive, would be beyond them.

There is no doubt that this has happened in some cases but the overall impression has to be that this has not occurred on any significant scale. We have already noted Charoen Pokphand's sale of its Lotus Supermarket chain to UK's Tesco and Standard Chartered acquisition of Thai Nakorthon Bank from the Wanglee family but these are the exception rather than the rule. As we come out of the crisis, it is quite remarkable how few Chinese corporations have fallen into the hands of western interests. Both the Bank of Thailand and IBRA have been trying to sell corporations that have passed into their hands through insolvency to western interests but so far with very limited success. Beyond the sale to Standard Chartered, Bank of Thailand's biggest successes have been in selling smaller banks and financial institutions to Singapore's DBS and UOB. IBRA's attempt to sell Bank Bali to Standard Chartered fell through when the British group encountered strong resistance from local management and staff to the takeover. The collapse of the sale brought the previous owners, the Ramly family back into the running for a buyback although they will almost certainly require a backer. Similarly IBRA's attempt to sell the Astra Conglomerate, whose major interest is assembling and distributing Toyota cars, to an American consortium also fell through, when local management and staff refused to co-operate in the latter's due dilligence enquiries.

In Hong Kong, Fuji Bank's attempt to sell its 51% but non-controlling interest in Kwong On Bank was constantly thwarted by the founding family's veto of any deal that didn't leave them with management control (*South China Morning Post*, November-December 1998, extensively). Fuji finally sold out to Singapore's DBS, which had agreed to the terms. In the Philippines, Lucio Tan has fought long and hard to retain control of Philippine Airlines. He has resisted all attempts by creditors, mostly western, to force him to withdraw from management in return for extended credit (*Far Eastern Economic Review*, 8 October 1998; 15 October 1998). An attempt by the Hong Kong-based Swire Group, which owns Cathay Pacific, to acquire the airline was also repulsed for similar reasons. At the time of writing, Tan appears to have been successful in retaining control and resisting all attempts by his creditors to oust him.

When Indonesian Chinese interests were forced to dispose of their US banking interests in East West (GT Group), United Commercial Bank (Liem) and California Security Bank (Raharja), to raise cash for survival, the sales were to American-Chinese and Taiwanese buyers (*Far Eastern Economic Review*, 27 August 1998; 24 September 1998).

The most spectacular successes by Chinese companies in retaining control in conditions of adversity were by the two largest Sino-Thai banks, Bangkok Bank and Thai Farmers Bank. Controlled by the Sophonpanich (Chin) and Lamsan families, both banks were buried under a mountain of bad debts by the crisis. Bangkok Bank was able to obtain some instant help from the Taiwanese Koo family (Chinatrust Bank) but such were its bad debts that it was assumed only massive recapitalisation from western sources could save them. In the event both families raised more than US$1 billion through new bond issues from domestic sources to recapitalize in the wake of the crisis. Domestic sources were happy to play a non-interventionist role and invest in the prospects and reputation of two of Thailand's leading Sino-Thai families. Both banks were recapitalized successfully leaving western institutions empty handed from the exercise.

A merger and acquisitions expert in Hong Kong recounts the saga of a large cashed-up American finance company that has been desparately trying to buy an Asian bank since the crisis began but despite lining up many potential deals has failed to clinch any of them. Business families will not relinquish control of their companies easily and even then, not without endeavouring to maintain some measure of interest. When interviewed some Chinese business people have claimed that retention of family control is particularly valuable in a crisis: 'it makes sense to put an owner's relatives in charge of a company in tough times, so hard decisions can be made....They had most to lose, so were probably best to run it' (Tom Kruesophon, *Far Eastern Economic Review*, 8 April 1999). The heads of some smaller Sino-Thai companies, who were surviving successfully, have described their tactics as a return to 'traditional values' including frugality, avoidance of debt, rapid expansion and diversification, family values and keeping outsiders out (Far Eastern Economic Review, 26 Feb 1998: 42-45).

Another route to survival was in finding new markets. Here the outstanding success story is Esprit, a fashion distributor. Originally, an American company, its world-wide interests were bought out by the Hong Kong branch when the US parent ran into difficulty. Despite being in the process of restructuring during the crisis, its retail turnover increased by 250% by turning to new markets in Europe; 43% of its sales are now in Germany with a further 20% in the rest of the EU (*Far Eastern Economic Review*, 15 April, 1999).

While it is easy to demonstrate the way Chinese business families have survived, the way the economy of the region has been restructured during the crisis is far more interesting. In this respect, it is by analysing what has occurred in Taiwan that we can most clearly see the massive changes that have taken place in the regional economy and the clear benefits of regionalisation and globalisation. While Taiwan has been a centre for the manufacture of computers, computer parts and peripherals for some time, mainly as OEM and ODM subcontractors for the US computer industry, the growing dominance of this sector, and the high-tech end of it in particular, is a phenomenon of the last few years. While Cathay Life Insurance controlled by the Tsai family retains its position as the largest privately-owned company, no less than seven of the ten largest companies are now in the high-tech sector. In 1995 there were only two.

What is of greater significance is that the ten leading high-tech companies in Taiwan all make the list of the top 30 corporations among the largest 500 privately-owned Chinese corporations in the whole East Asian region. While this trend is the most advanced in Taiwan, it is happening on a smaller scale throughout the region. No longer can it be claimed that the diaspora Chinese economy is dominated by the property, finance and gambling sectors as Japanese analyst Yoshihara had earlier claimed in *The Rise of Ersatz Capitalism in Southeast Asia* (Yoshihara, 1988).

Table 9.4 shows that in 1999, 27 of the largest 100 companies region-wide were in high-tech or telecommunications against 14 in propery development. In 1995, there were only six against 23 property developers. Another indication of this change is that the three new Chinese billionaires, who have forced their way into the 100 richest business families in Asia in 2000, are all in the high-tech sector: Patrick Wang (Johnson Electric) and Richard Li (Pacific Cyberworks) from Hong Kong and Barry Lam (Quanta Computers) from Taiwan. At the same time a number of essentially property-centred entrepreneurs have dropped out despite the recovery in property markets in 1999.

**Table 9.4 Industrial Composition of 100 Largest Chinese-owned
Private Companies 1999**

	1999	1995	Change 1995-99
Conglomerates	7	8	(1)
Property Developers	14	23	(9)
Finance & Banking	16	20	(4)
High-tech & Telecom	27	6	21
Manufacturing	23	26	(3)
Public Utilities	4	4	-
Media	2	2	-
Gambling & Leisure	4	8	(4)
Transport & Shipping	1	1	-
Retail	2	2	-

Source: Yazhou Zhoukan.

In much the same way as new sectors have come through or consolidated their positions in previous major economic crises, so the current crisis has seen the rise of the inter-related high-tech, internet, telecommunications, software and computer sectors as a substantial core of the new regional economy. This has happened by the conscious design of the entrepreneurs involved, who have both spotted openings in the new economy and taken advantage of them, and because the crisis has put pressure on the more traditional sectors like property and finance, both of which were very badly affectd by the crisis.

Taiwan's links with the American computer industry have paid off in the new economy. Taiwanese companies have long acted as sub-contractors for the American giants. Beginning as OEM manufacturers, they progressed to ODM and now finally an increasing number have become OBM. The dependence of the US industry on its Taiwanese contractors was exposed when Taiwan was hit by a massive earthquake in September 1999, which halted component manufacture and quickly put pressure on the US companies' own production and distribution systems. The links with the US industry also paid off in that export markets remained open and even increased during the worst period of the crisis, while regional markets for other products seized up for lack of demand, as the credit squeeze slowed and even halted business activity.

However, seeing the growth of computer-related industry in Taiwan as essentially the product of the US industry's subcontracting is to misunderstand what is happening. There is no doubt that this was, is and is likely to remain important but it is only part of the picture. There are also other forces at work. Many Taiwanese choose to study in the United States, this is particularly so for post-graduate work in science and technology. Gaining their qualifications in the United States, many stay on to serve their apprenticeships with US corporations. Increasing numbers have served their time in Silicon Valley[5]. They have finally returned to Taiwan in many cases to set up their own businesses and become entrepreneurs. These returnees have been critical among the large number of new set-ups in software and internet companies. The Taiwanese 'high-tech revolution' is not confined to the hardware but is increasingly focussed on new high-tech sectors in software, internet and the new sciences like biotechnology. It is Taiwan's edge in these new technologies, which is giving it an advantage over Japan in the new economy that it could never have hoped to gain in the old economy dominated by industrial technology.

Noteworthy in this rise of the high-tech sector was the high rate of profitability of these companies. In stark contrast to many NASDAQ stocks, only one of the ten leading companies failed to return a profit in 1999, with half showing increases in profitability in excess of 40% (*Yazhou Zhoukan*, 1 November 1999).

It would be wrong, however, to see the move in Taiwan into high-tech development as resulting purely from the crisis. It is clear that such changes were already underway before the crisis began. The crisis, however, introduced a sense of urgency that pushed the sector forward. At the same time older sectors both in finance and manufacturing were badly affected by the crisis. These two factors combined to enable the high-tech sector to consolidate itself as the new core of the Taiwanese economy and the key indicator of the way the economy needed to develop in the future. In much the same way as the Great Depression established autos, aircraft, petrochemicals and the new consumer-oriented industries as critical cores of the new US economy, so this crisis enabled the computer sector to firmly establish itself as the key to Taiwan's economic future.

What is also noticeable about Taiwanese companies during the crisis was the way many entrepreneurs used the crisis to restructure their production facilities and seek to win new long-term markets. A significant indicator of this is the new production facilities established by a number of Taiwanese companies, including Acer, in Dongguan in Guangdong Province in China. Dongguan was previously a centre of the so-called processing trade wherein Hong Kong companies part-processed labour-intensive manufacturing activities there. The county has been transformed in the last three years by the influx of Taiwanese computer companies, who are both manufacturing for export and with a longer-term view of supplying the China market. With industrial labour readily available this considerably reduces manufacturing costs. Lower labour costs are, however, only part of the advantages available. With nearby Shenzhen rapidly establishing itself as a centre of high-tech development in China, the possibility of establishing some research and development facilities utilising underemployed Chinese engineers from research institutes in the northern and inland provinces must be attractive. There are some indications that this is already underway but this needs to be followed up with further research.

What has happened in Taiwan is happening elsewhere even if at a less frantic pace. Hong Kong, long seen as heavily overdependent on the property and stock markets, despite its position as a centre of regional finance and as the management and marketing hub for its labour-intensive manufacturing industries in Guangdong, has seen an explosion of internet related activity. Leading the charge is Richard Li's Pacific Century Cyberworks. It's rapid expansion has propelled it into the ranks of the largest corporations in Hong Kong in very rapid time. Capitalised at more than US$3 billion by the end of 1999, its growth to date has been fuelled largely by acquisitions rather than its own technological development. The very fact that it has been able to expand so rapidly in this fashion, however, indicates the changes that are taking place. Three of its most important acquisitions in 1999 were Hong Kong-based: Outblaze, an email service provider, Link, an internet data host, and Netcel, an ecommerce company. Pacific Century Cyberworks has targetted acquisitions that will quickly generate content or provide distribution channels for the superfast internet service it will launch in 2000, utilising a combination of satellite and local cable-TV links (*Far Eastern Economic Review*, 20 January, 2000)[6].

Li is the younger son of Hong Kong tycoon, Li Ka-shing, and his family's property links and financial backing were probably what helped him gain the contract to develop the Hong Kong government's high-tech brainchild, Cyberport[7]. The latter is intended to be a high-tech special zone enabling Hong Kong to attract both the global and local companies that it hopes will propel it towards becoming a regional centre of the new computer-based economy.

However, the changes are not limited to the new start-ups. Long-established staid property tycoons are moving into the computer age. The

Kwok family's Sun Hung Kai Property, previously seen as a pure property play, signed a deal with Microsoft to develop a system to wire all its buildings for the new technology. They have also established a technology company, Sunevision, which they intend to capitalise with a US$1 billion IPO in Hong Kong in 2000 (*Far Eastern Economic Review*, 24 February, 2000). Acquisitions and buy-ins are anticipated. Ng Teng-fong's Sino Land quickly followed suit. Tycoon Lee Shau Kee of Henderson Land has entered a partnership with David Li's Bank of East Asia and another Hong Kong entrepreneur, Gus Chow of Harmony Assets, to develop an e-commerce business using the customer bases of the bank and his own Hong Kong and China Gas (*South China Morning Post*, 19 February, 2000:B3).

Even the largest of Hong Kong's tycoons, Li Ka-shing is finding that telecommunications, particularly his mobile phone operations, is the fastest growing sector of his business interests. Despite being the largest operator of container ports in the world, with interest in China, Rotterdam, Felixstowe in the UK and at both ends of the Panama Canal, and his substantial regional property interests, it is telecommunications that is increasingly driving his business. With his younger son, Richard, now the leading internet entrepreneur in Hong Kong, the shift in the group's centre of gravity is becoming clear.

The same trends can be seen in Singapore. The city state's disadvantage, however, is the very tight rein the government has kept on economic development in the past. The result is that there isn't the same entreprenurial zeal amongst its inhabitants, particularly where risky ventures are involved. The most important new technology company is Creative Technology. Unlike Pacific Century Cyberworks, Creative Technology started with a core business producing audio and video components. It is reputed to be the world's largest manufacturer of soundcards. However, like Pacific Century Cyberworks it is now concentrating on taking stakes in high-tech start-ups which can be related to its core business. In 1999, among 20 part-acquisitions, it bought into a internet telephony provider, an e-commerce company selling audio products and an e-music business that enables music to be downloaded into one of its own products.

The fact that Creative were able to buy into so many of these new companies is an indication of where new developments are taking place and how quickly new Singapore-based entrepreneurs are joining the new computer-based economy now that the government has acknowledged the need for innovation and encouragement of its entrepreneurs. One of these is Pacific Internet. Already a provider of internet services, it is planning to follow Pacific Century Cyberworks in seeking to become a major regional internet player.

Despite the impact of the crisis on Thailand, high-tech and telecommunications companies have survived and their survival has shifted the balance of Chinese business. The Sino-Thai business sector was long dominated by the banking sector, by agribusiness and by property developers. With all of these suffering badly, the balance has changed. While three banks (Bangkok Bank, Thai Farmers Bank and Bank of Ayudhya) and one property developer (Italian-Thai Development) remain among the ten largest Sino-Thai companies, there are also four in telecommunication and two in advanced electronics.

As this was at least initially a financial crisis, it is interesting to see what has happened to the region's Chinese banks. The Chinese-owned banks in Singapore have firmly established themselves as the strongest of the regional banks excepting the British-controlled and managed HSBC and possibly Citibank[8]. The list of the largest Chinese banks puts three Singapore-based banks at the head of the list (OCBC, UOB and OUB). Next came the Koo family's Chinatrust from Taiwan, closely followed by the Hong Kong Li family's Bank of East Asia, the Bangkok-based Chin group's Bangkok Bank and the Hong Kong-based, but controlled from Malaysia by the Quek family, Dao Heng. These were followed by the Ooi family's Bank Internasional Indonesia, the Ty's Metrobank from Manila and the Lamsan's Thai Farmers Bank.

It is important to note, however, that substantial Chinese banks in Indonesia, most notably BCA, and Bangkok Metropolitan Bank have gone under. Another long-established Chinese bank in Bangkok, Thai Nakorthon Bank, has been sold to British interests. In addition, all Chinese-owned banks are threatened with unwanted mergers in Malaysia following the govenment's decision to restructure the whole financial sector. Only the Teh family's Public Bank is likely to emerge from this strengthened. A number of banks in Taiwan also faltered during the crisis. The impact of this is that the Chinese financial sector is increasingly centred in Hong Kong and Singapore strengthening an already established pattern. In this respect, it is important to note that while only two Hong Kong-based banks made the top ten, a number of smaller banks, of which Wing Lung is probably the most important, not only survived but prospered. There is greater depth in the banking sector in Hong Kong than in Singapore with a number of quite vibrant smaller banks.

Conclusion

The first conclusion we have to draw is that Chinese Diaspora capitalism is alive and well. Some sectors are very badly bruised, those heavily concentrated in the worst affected economies, Indonesia, Malaysia and Thailand and the property sector everywhere have taken a pumelling. This has produced many failures and has so weakened others that they will take a long time to recover. Overall, however, the majority of even the largest and most exposed have survived. At the same time, many new entrants, including some who have achieved tycoon status already, have emerged in the high-tech sector. The crisis has seen a cycle of decline and regeneration which has transformed the regional economy.

The really big shift, however, is the dramatic restructuring that has taken place in the Chinese economy. Starting in Taiwan and spreading out from there, the new image of the Chinese diaspora economy is of a substantial shift towards computer-based and high-tech sectors. Taiwan has clearly established itself as the major centre of the technology but Hong Kong and Singapore are rapidly following although less in developing the technology than in using it the most effectively. Where all three Chinese mini-entities seem to have excelled is in the large number of new start-ups seeking to exploit the new computer and internet-based economy. Nothing illustrates the new economy and its growing confidence more than Richard Li's Pacific Century Cyberworks' 'outrageous' bid for Hong Kong Telecom, majority-owned by the UK's Cable and Wireless Group. Bidding for a company several times the size of its own and in opposition to state-owned SingTel's bid confirms the arrival of the new economy and its entrepreneurs. When asked why he thought he could bid for HK Telecom against SingTel, Li replied that he woke up one morning and thought 'I can do more with this company than they can' (*Far Eastern Economic Review*, 24 February, 2000). Entrepreneurship is still the critical variable carrying Chinese business forward. Despite the shocks inflicted by the crisis, the arrival of new entrepreneurs like Richard Li, Patrick Wang and Barry Lam during the crisis and the new directions they are both taking and forcing the regional economy towards suggest that the crisis has begun to regenerate the Chinese business sector throughout the region. Expect a new challenge to US dominance of this sector from this direction.

Notes

[1] The majority of FDI into China from 1985 onwards came from Chinese sources, principally Hong Kong and Taiwan. In the same period half of China's exports passed through Hong Kong marketing channels on their way into world markets (Lever-Tracy et al, 1996).

[2] The United States was by far the most seriously affected of the major economies. Industrial production fell by 50% between 1929 and 1933 and was barely two-thirds of the 1929 figure in 1939. The US remained in serious recession until the outbreak of the Pacific War in 1941 after which war production soon revived the economy.

[3] The various stock exchange and corporate handbooks produced throughout the region clearly show the close relationship between principal shareholders and the top executive functions.

[4] This term 'Chinese business sphere' was first coined by Masaaki Yamaguchi (1993) to delineate a region in which the majority of private business transactions passed through Chinese business networks.

[5] It is reported that almost half the computer engineers in Silicon Valley are of either Chinese or Indian ethnicity. The advance of both the Indian and Chinese diaspora economies in the high-tech areas might be seen as, at least in part, linked to this.

[6] Initially, it looked as though Pacific Century Cyberworks was going to be a Hong Kong version of Masayoshi Son's Softbank, essentially an investor in high-tech stocks, however, these subsequent developments have suggested that was a premature and erroneous judgement.

[7] With his elder brother Victor being groomed to take over the family's core group (Hutchinson Whampoa, Cheung Kong and Hong Kong Electric), Richard seems to have decided or been encouraged to strike out on his own.

[8] A number of state-owned banks, particularly DBS in Singapore and Maybank in Malaysia, are perhaps larger than the private banks, but they do not have the regional coverage. DBS is beginning to internationalise its operations but it still lags behind OCBC and UOB, and possibly even OUB, in this respect.

References

Clegg, S. and Redding, S.G. (eds) (1990), *Capitalism in Contrasting Cultures*, Walter de Gruyter, Berlin.

Far Eastern Economic Review, various dates.

Forbes, 5 July, 1999, 'The World's Working Rich'; 28 July 1997, 'Asia at a crossroads'.

Fortune, 'The World's Billionaires' (annual list).

Greenhalgh, S. (1988), 'Families and Networks in Taiwan's Economic Development' in Winckler, E. and Greenhalgh (eds), *Contending Approaches to the Political Economy of Taiwan*, An East Gate Book, Armonk.

Hamilton, G. (ed) (1991), *Business Networks and Economic Development in East and Southeast Asia*, University of Hong Kong Press, Hong Kong.

Kunio, K. (1988), *The Rise of Ersatz Capitalism in Southeast Asia*, Oxford University Press, Singapore.

Lever-Tracy, C., Ip, D. and Tracy, N. (1996), *The Chinese Diaspora and Mainland China: A New Economic Synergy*, Macmillan, Houndmills.

Redding, S.G. (1990), *The Spirit of Chinese Capitalism*, Walter de Gruyter, Berlin.
Redding, S.G. (1991), 'Weak Organisations and Strong Linkages: Managerial Ideology and Chinese Family Business Networks', in Hamilton, G. (ed), *Business Networks and Economic Development in East and Southeast Asia*, Hong Kong University Press, Hong Kong.
Sit, V.F.S. and Wong, S.L. (1989), *Small and Medium Industries in an Export-Oriented Economy: The Case of Hong Kong*, Hong Kong University Press, Hong Kong.
South China Morning Post, various dates.
The Australian, 19 April, 1999, 'The top 100 companies in 10 Asian Markets'.
The Australian, 27 January, 2000, 'Asian Billionaires in 2000: the 100 wealthiest families'.
Winkler, E. and Greenhalgh, S. (eds), *Contending Approaches to the Political Economy of Taiwan*, East Gate, Armonk.
Yazhou Zhoukan (The Chinese 500 is contained in the issue published in the first week in November each year).

10   The Irrelevance of Japan

CONSTANCE LEVER-TRACY

Introduction

How relevant is Japanese leadership to the process of Asian recovery from the crisis? One of the most repeated and least examined assertions by pundits has been that, until the Japanese economy recovered, there would be no hope for the rest of Asia pulling out of the crisis[1]. It is necessary to question this taken-for-granted maxim both because it seems to have little foundation, and because, unless this fallacy is cleared out of the way, the examination of the initiatives and prospects of the diaspora Chinese in the region, might seem a diversion from the real game.

In Chapter 1 we pointed to the surprisingly low visibility of diaspora Chinese capital in most academic writings about the prospects for Asian development and the unfolding of the 'Asian Miracle'. Too often the towering shadow of Japan has blocked the view. The networks of small and medium Chinese family firms were for long dismissed as relics of an archaic, traditional pariah or middlemen minority form of capitalism, soon to be displaced as the societies modernised. Even when rising tycoons began to feature prominently in such places as the Forbes list of the world's wealthiest billionaires, this could seem like the trivia of society gossip, given the absence of their firms from supposedly more serious rollcalls of the world's large corporations, such as the Fortune 500. Even the biggest Chinese conglomerates, which may involve an accumulation of hundreds of small and medium companies in the ownership of one family, did not register in a comparison of their individual components with the corporate behemoths of Japan, although the aggregate workforce, turnover or profits of the wealthiest might be of a comparable order of magnitude. The fact that Chinese capitalists were based in a multitude of states, often as minorities, and that their transnational activities were linked by informal networks rather than bureaucratic hierarchies, also made it difficult to notice their emergence as significant global actors, especially when the attention of observers was distracted elsewhere.

Most of the focus on the region has rather been on the high profile activities of developmental states and of multinational, particularly

Japanese corporations, whose projects and achievements were readily accessible in policy documents, company reports and national statistics and were immediately visible in familiar logos and brand names. A stream of writings have argued the lead role of Japan in Asia's prospective and actual economic development. Again and again the 'imminent economic hegemony' of Japan in the region, always just around the corner, has been predicted. In 1984 Kojima and Ozawa's *Japan's General Trading Companies: Merchants of Economic Development* celebrated the role of Japan as leader of rising Asia's 'flight of geese'. The following year Abbeglen and Stalk wrote of there being an 'Asian option' to Japan's previously US focussed policies (1985, p. 242). He argued that Japan had provided the model, the technology, the capital and the markets for the expanding economies of Asia and that while other markets and investors were important 'it is interaction with Japan that has made them remarkable' (1985, p. 262). Others have reiterated the thesis, more often with dire warnings of permanent subordination and exploitation than with praise (Rob, 1990; Ping 1990; Cronin, 1991; Morris-Suzuki, 1992). As late as 1996 Hatch and Yamamura's luridly titled *Asia in Japan's Embrace* claimed that Asia as an 'American lake' had all but dried up and a 'Japanese lake' was rising in its place.

> Japanese capital and technology are stitching together the disparate economies of Asia, integrating them into a multilevel production alliance (p. x).

If Japan had indeed been the initiator, and the continuing leader of the previous rapid growth in the region, her indispensability in any recovery would seem a safe assertion. Yet, as this chapter will seek to demonstrate, the central initiating role of Japan in Asia's original take off may be called into question, her continuing role in the 1980s has been much overstated and for most of the 1990s, prior to the crisis, while East and Southeast Asia and China experienced accelerating growth, the Japanese economy was stagnating and its relative regional economic power was in sharp decline.

The claims of Japanese primacy have been made possible by an analysis which focuses on Japanese trade and investment, ignoring their importance relative to the contribution of others. The impact of Japanese multinationals in the fate of the region is discussed, while little cognisance is taken of local capitalist forces. Breathless accounts of the rising might of Japanese presence in the region are rarely accompanied by any indication of the scale and growth of the total economy, of which they have actually formed a relatively small and recently shrinking portion. It is as if someone was trying to construct a jigsaw picture using only one kind of pieces, with no information about the nature or even the quantity of the missing parts, or the size of the overall picture.

Japan as the Starting Motor?

The Take Off

Writers such as Kojima and Ozawa (1984), Arrighi et al (1993), and Ozawa (1993) have argued that it was the inflows of manufacturing investment into the East Asian NIEs by (mainly small and medium) Japanese firms, from the mid 1960s, that triggered their industrialisation in the first place. Japan is presented as the initiator of export oriented manufacturing in East and later Southeast Asia. As she moved up the technology ladder herself, she handed down, initially to the first rank of 'follower geese' the less skilled industries she was leaving behind. The followers were trained and equipped through Japanese subsidiaries and contractors, and their products were given access to world markets by the *sogo shosha*. the major Japanese trading companies. As each line of geese moved up, the processes it was relinquishing were passed on to those further behind.

While the account has an appealing symmetry, the evidence cited by its proponents is not fully convincing. Too rapid a jump seems to be made from the fact of succession to the presumption of a transplantation involving a symbiotic or subcontracting relationship initiated and overseen by the Japanese. Yet the scale of Japanese investment, manufacturing and exporting activities in East Asia, relative to non Japanese activity, seems to have always been too small for the consequences attributed to it and the major intermediary role claimed for the *sogo shosha* seems undemonstrated and unlikely. As a result of the hypothesis, the possible initiating role of other local players is left unexplored.

Kojima and Ozawa cite a total of 1,128 small and medium Japanese manufacturing projects offshored to all parts of Asia by the end of 1975 (after which they admit the influx slowed down). The smallness of this becomes apparent when we contrast it with the over 90,000 manufacturing establishments in Taiwan by 1981 (Orru, 1991, p. 10) or the nearly 11,000 small firms, just in garments and electronics, in HK in 1980 (Chiu & Lui, 1995). Evans (1987, pp. 206-209) has pointed out the very limited role of FDI in the development of the NIE, pointing out that even in

1964 domestic savings in Taiwan accounted for 65% of capital formation, rising to 95% by 1969 and that in 1987 the FDI stock in South Korea was minimal, worth only 2% of that in Brazil.

An interesting but long ignored article by Hone (1974) argued that for the period up to 1972, the role of all foreign direct investment (FDI), in East Asia's export performance had been much exaggerated. He claimed that already between 1962 and 1969 non Japan Asia's manufacturing exports had risen from 47% to 60% of all developing countries' manufacturing exports, but that little of this rise was due to FDI from any foreign source. In 1972 all FDI ventures in HK employed only 13% of the total manufacturing labour force and accounted for only 10% of domestic exports (p. 147). In Korea in 1972 accumulated Japanese investment was worth $60 million. In that year manufacturing exports were worth $750 million. He estimated that by 1972 all foreign invested companies accounted for no more than 10% of East Asia's exports, with the recent inflows of Japanese investments in Taiwan, Korea and Singapore accounting for under 5% of their domestic exports.

Some of the misconception, he argued, came from too narrow a focus on the highly visible export processing zones. For example, Japanese and US firms were significant investors in the Kaohsiung zone in Taiwan, but in 1970 exports from the zone were only $109 million out of total manufacturing exports from the country of $800 million. Two years later the proportion had fallen to $300 million out of a total of $2.7 billion. (p.146). Nearly all exports from Taiwan came from domestically owned firms, most in small towns and in the countryside where few foreign firms penetrated.

Hone argued that the real foreign impact came, not through investment, but through the buying and commissioning power of major oligopolistic Western and Japanese retailers and buying groups. Yet here, it seems clear to us, the Japanese role was and has continued to be limited by the persisting low level of their purchases of manufactured goods from Asia, which was primarily a source of their raw materials. The only significant contribution of Japanese buyers could therefore have been as intermediaries for sales to the West. But given the presence in the region of direct agents of the Western retailers, as well as tens of thousands of small, local import, export agents and the Chinese networks extending into America itself through students and emigration (Greenhalgh, 1988), it is hard to see what long term justification there might have been for such a Japanese middleman role here.

In fact Kojima and Ozawa (1984 p. 37) give figures indicating that the major *sogo shosha* were primarily involved in trade to and from Japan and to only a very limited extent in trade between third countries. It is estimated that in the 1970s, while a third of Taiwan's trade with Japan was through the *sogo shosha,* only 15% of her trade with the rest of the world was thus mediated[2]. Indeed, surprisingly, the boot may be on the other foot. Kagatoni's research (1998) suggests that it was the Kobe community of overseas Chinese traders, during the 1930s, who through their networks were able to distribute Japanese cotton textiles, one of the main export products in pre-war Japan, in Asian markets.

Global Power but not Regional Hegemon

It was during the 1980s that Japan emerged as a major global economic power, rivalling the West in exports and foreign investment. Yet even now the weight of Japanese investment and trade in the region was far from being at hegemonic levels. Furthermore, although rising absolutely, both had actually been generally shrinking or constant in relative terms. The region was playing a declining part in Japan's overseas investment strategies, which increasingly directed its FDI towards its major markets in the West. More importantly, Japan was contributing a smaller proportion to the rising foreign investment flows entering the rapidly developing countries of the region, especially in the latter half of the 1980s. Intra-regional investment, particularly from Hong Kong[3] and Taiwan was running neck and neck, or actually overtaking Japan, by the start of the 1990s (Lever-Tracy and Tracy, 1993)[4].

Despite a rapid absolute growth in trade between Japan and the region, there was nothing in relative trade figures to indicate either an increasing emphasis by Japan on the region, or any tendency for a rising Japanese role in integrating the region through trade. Japan's contribution as a purchaser of the region's products was declining. Her role as a supplier was also becoming relatively less important for the NIEs and was increasing only slightly for the ASEAN countries. Meanwhile trade between the countries of the region was increasing both absolutely and relatively.

Before the Crisis - The Decline of Japan in Asia

In the 1990s, although Japan's interest in the region, relative to other parts of the world, seems to have increased again, the scale of her overall new investment flows abroad have collapsed, while the inflows to the region from other sources have continued to rise. While Japan's trade with the region continued to grow, the region's own productive and effective demand capacities grew much faster than those of Japan. Calculating how the intersection of several contrary trends has affected Japan's relative weight in the region is by no means straightforward. The task is further complicated (especially where investment is concerned) by the unsatisfactory or delayed nature of much relevant data. Few recent publications in the field have taken due cognisance of the extent to which the balance has been changing, and thus the aim of this chapter is to attempt a timely sketch of the changes in the regional weight of Japan's investment and trade in the years leading up to the crisis.

Sources of Investment

Table 10.1 shows the changing average annual size of outward flows of world foreign investment between 1982 and 1987 and between 1988 and 1991 and annually from 1992 to 1996. It also shows the changing relative importance as foreign investors of the countries of East Asia and of Japan in each of the three periods. The last column shows a pattern of continuous growth in the value of world FDI over the three periods. Total average annual world flows were three times higher in the second period than in the first. In each of the following four years the world total rose again, although much more slowly than before, to reach an estimated US$352.5 billion in 1995 with some fall back to US$333.6 billion in 1996.

What is most startling in this table is the absolute and relative recent decline of Japanese investment. In the second period Japan had been investing on average over US$42 billion a year, over four and a half times more than in the first period. Rising faster than the world total, their proportion of world investment had increased from 13.4% to 20.1%. In the subsequent five years, however, the value of their annual investment has fallen to less than half this figure and their relative position has fallen even further, to around 7% of the world total (less than it had been in the first period, in the earlier 1980s).

In contrast, the joint value of foreign investment undertaken by the NIEs, the ASEAN countries and China rose between each of the three periods. Their combined total, worth little more than 1.4% of world direct foreign investment in the first period had risen to 5.2% in the second and to over 14% by 1996. Over the whole of the third period their combined total was more than one and a half times larger than that of Japan and their proportion of world investment outflows was over 50% greater than Japan's. The value of their new investment flows exceeded those of Japan every year after 1992.

Table 10.1 Distribution of Direct Foreign Investment Outflows by Source (Percent of world total outflows. US$bn in brackets)

	4 NIEs	Other ASEAN	China	**Sub Total**	Japan	World
1982-1987 Annual Average	0.9		0.5	1.4	13.4	100 (67.9)
1988 – 1991 Annual Average	4.2	0.6	0.4	5.2	20.1	100 (210.4)
1992	6.3	0.4	2.0	8.7	8.7	100 (200.8)
1993	9.8	0.9	1.8	12.5	5.7	100 (240.9)
1994	10.6	1.1	0.7	12.4	6.4	100 (284.3)
1995	10.1	1.2	0.6	11.8	6.4	100 (352.5)
1996	11.9	1.6	0.6	14.1	7.0	100 (333.6)

Sources: Figures have been taken from various editions of WIR. Since figures for earlier years are corrected retrospectively each year, the latest edition has been used where possible. Earlier figures for particular periods may, however, not be available from these later editions. The table thus sometimes combines data from different editions.

The scale of recent decline in investment flows from Japan has been sufficient to reverse the previous tendency for their share of accumulated outward stocks of world foreign investment to grow. As Table 10.2 indicates these went up from 3.7% in 1980 to 6.5% in 1985 and to 12.2% in 1990. By 1994 this had fallen to under 12% and by 1995 it was estimated to be little over 11%. The proportion accounted for by Europe and America also fell slightly in the 1990s, leaving East Asia as the clear beneficiary. It is notable that East Asia's proportion of the total world stocks, at 6% in 1995, is nearly as great as Japan's 6.5% in 1985, when its significance as a core capitalist power was already widely recognised[5].

Destinations of Investment

Overall data for inward and outward investment flows are reasonably available in a timely fashion. Reliable breakdowns of a particular country's outward flows by destination and of particular inward flows by source are harder to obtain. It is thus much harder to ascertain how much of Japan's FDI is going to East Asia and how much of East Asia's inflows are coming from Japan.

For Japanese investment, for example, the generally cited destination figures are based on notification of investment intentions rather than on actual flows. Table 10.3 makes clear how these overestimate the actual flows and how they do so with little consistency from one year or one period to another. For the first of our three periods, 1982-1987, actual total outflows from Japan were only equal to 58% of the notifications made during this period. In the second period, up to 1991, when the size of actual outflows was much larger, these represented nearly 80% of notifications. In the third period, however, actual outflows shrank, progressively, by much more than did the notifications, totalling less than half the value of notifications for the period. Relying on notifications data leads to serious overstatement of the value of Japanese FDI and to underestimation of the extent of its decline in the 1990s. Direct figures on destinations are, however only available in this way.

The best we can do perhaps is to take the notifications proportions as an indication of the distribution of Japanese investment **intentions**, but not of their realised value. Column a) of table 10.4a shows a rapid and continuing increase in the attractiveness of East Asian destinations to Japanese investors, from 10% of all notifications in 1990 to 24% in 1995[6]. If we multiply these proportions of notifications to E Asia by the total of actual Japanese outflows, we obtain in column c) a notional $ value for investment flows from Japan to East Asia. While 'notional' these figures are certainly closer to reality than those provided by notification data alone.

It is clear the rising interest in East Asia has only partly countered the decline in total investment outflows. Sums (notional) directed to Asia for 1992-1994 are lower than for 1990 and 1991, although the 1990 figure of US$5 billion is again overtaken in 1995.

Table 10.2 Outward FDI Stock. Share of World Total

	1980	1985	1990	1994	1995*
Japan	3.7	6.5	12.2	11.8	11.2
Dev Asia	0.3	0.8	2.3	5.1	6.0
USA	42.9	36.6	25.8	25.3	25.8
Europe	46.0	45.6	50.7	49.4	48.8
Sub total	88.9	82.2	76.5	74.7	74.6
World	100	100	100	100	100

* Estimate

Source: WIR, various dates.

Table 10.3 Japanese Foreign Direct Investment to All destinations. Notifications and Actual Outflows (annual averages - US$bn)

	a) Notification	b) Outflows	c) b as % of a
1982 – 1987	15.6	9.1	58
1988 – 1991	53.3	42.2	79
1992	34.1	21.9	64
1993	36.0	15.5	43
1994	41.0	18.5	45
1995	50.6	21.3	42
1992 – 1995	40.4	19.3	48

Sources: JSY (various dates); WIR (1993, 1996); data on notifications for 1994, 1995 supplied by Japanese Embassy.

A more crucial question is about the relative importance of Japanese investment to recipient countries. What proportion is Japanese investment of their total inflows? For this one must go to the data collected by host countries. This too is often a count of notifications or approvals at the receiving end, and is hard to combine as different countries express this in different currencies and according to different criteria. This information is also usually only available with a considerable time lag. Table 10.4b seeks to avoid these difficulties and speed up the process by taking the 'notional' figures of Japanese investment in East Asia which have been calculated in table 10.4a, and expressing them as a proportion of the total inflows into East Asia. This is shown in table 10.4b, column c). The proportion of inflows that may be presumed to have come from Japan in 1990-1991 was 26%, a substantial although far from hegemonic figure. For 1992-1995, however, this has fallen to a quite marginal 8%, with no evidence of recovery on the way.

The huge and rapidly growing worldwide outflows of Japanese investment in the later 1980s and early 1990s also caught the attention of academic writers, and they were integrated into conceptual schema of imminent regional hegemony and Japanese domination of integrated regional production networks. Few stopped to calculate their relative weight and to note that because of their redirection towards the West and the rapid growth of the region, they actually represented a declining proportion of total foreign investment in the region. Since 1992 Japanese investment in the region has fallen both absolutely and relatively, particularly as compared with cross investments from within the region itself, despite the indications that its intentions are increasingly focussed here. While Japan, as one of the world's great economic powers, is of a course a significant foreign investor in the region, it is not now, nor has it in the past been, a dominant one. It is only by disregard of the other local actors that it may appear so.

Japan's Trade with the Region

Trade with the countries of the region has been of growing importance to Japan in the 1990s. In 1990 they bought 23.6% of her exports, a figure which surged to 35.6% by 1996. This trend applied to both the NIEs and to the countries of ASEAN. While Japan's sales to China had remained a low and static proportion through the 1980s it too was rising fast over the 1990s. The role of these countries together as suppliers of Japan's imports had also grown, from providing 21.8% in 1989 to 31.2% in 1996 although the pattern was less consistent between countries, with lower proportions from the ASEAN countries and particularly large increases from China.

Table 10.4 Japanese Foreign Direct Investment (FDI) in East Asia

a) Value

	a) FDI Notification to Asia as % of all Notifications	b) Actual FDI flows from Japan to all destinations US$bn	c) $b \times a$ Notional flow to Asia US$bn
1990	10.4	48.1	5.0
1991	12.0	42.6	5.1
1992	16.1	21.9	3.5
1993	16.0	15.5	2.5
1994	22.8	18.5	4.2
1995	24.0*	21.3	5.1
1990-91	-	90.7	10.1
1992-95	-	77.2	15.3

* Includes other parts of Asia as well.

b) As proportion of inflows to host country

	a) FDI inflows to E. Asia from all sources US$bn	b) Notional flows from Japan to E. Asia US$bn	c) b as % of a
1990	19.3	5.0	26
1991	20.3	5.1	25
1992	26.5	3.5	13
1993	45.5	2.5	5
1994	50.7	4.2	8
1995	62.0	5.1	8
1990-1991	39.6	10.1	26
1992-1995	185.0	15.3	8

Sources: *WIR* various years.

Overall, then, the countries of the region (particularly China) were of growing importance to Japan as customers and also as suppliers. A crisis in the region would be of major concern to Japan. Nonetheless Japan's trade outside the region, especially with America and Europe, remained of primary importance to her.

When we look at this trade through the other end of the telescope, from the point of view of the countries of the region, it becomes clear trade with Japan has been relatively less important to them than it has been to Japan. While over a third of Japan's exports went to the region in 1996, these constituted only 18.2% of the region's imports, a proportion that had declined from 20.5% in 1990. Japan was even less important as a buyer of the region's growing exports, accounting for only 13.9% of these in 1996, down from 14.5% in 1990. This fall was greater for the ASEAN countries and would have been even larger if China had been excluded from the picture (*DOTS annual*, 1997).

It is clear, then, that on the eve of the crisis East Asia had become of greater importance to Japan, as supplier and especially as customer, than was Japan to East Asia. It is also clear that the countries of the region were, collectively, more important to each other, as suppliers and as customers, than was Japan. By 1996 imports from other countries of the region were worth two and a half times more than their imports from Japan and exports to these countries were worth 3.4 times more than their exports to Japan. Although trade between Japan and East Asia had not collapsed in recent years, in the way that investment flows had, neither their level nor the trends supported a case for an actual or imminent hegemony of Japan over the region'. Indeed it seems more logical to ask whether a recovery in the rest of Asia can pull the Japanese economy out of its stagnation, than to seek in Japan a recovery motor for the rest of Asia.

Icebergs?

Objections

This chapter has focussed on broad patterns of investment and trade and the relative weights of different actors. Those who have argued for the rising power of Japan in the region have rarely given much space to such data but they do periodically caution against attaching too much importance to them. The argument is that broad quantitative data underestimate the less visible Japanese influence for a number of reasons.

The first reason suggested is that sums invested do not give an adequate picture of the impact of Japanese FDI because affiliates are often joint ventures with relatively low Japanese equity, Japanese inputs are often in the form of locally raised loans rather than equity (Tokunaga, 1992b) and because Japanese affiliates are just the tip of an iceberg of subcontractors (Ozawa, 1993; Arrighi, 1993).

The second reason proposed is that Japanese power and profits are enforced through their control over the supply of higher tech components and capital equipment (Bernard and Ravenshill 1995; Morris-Suzuki, 1992).

The third, and perhaps most important argument is that Japanese act as co-ordinators of regional production networks in which local producers and even nation states are just bit players with no capacity to affect the overall picture (Bernard and Ravenshill, 1995: Morris-Suzuki, 1992; Tokunaga, 1992a).

Responses

These arguments can only be answered very briefly here. Their major weakness is that they are scarcely ever quantified in a way that would enable them to bear the weight of the case that is resting on them. While some are valid, although insufficient as evidence, others are based on undemonstrated assumptions, for example about the effective transfer of the Japanese subcontracting system to its overseas affiliates and about the effective implementation of what, one might suspect, are often utopian schemes by Japanese government bodies for regional co-ordination. They are also based quite often on claims about the impossibility of technological catching up by the East Asians (Bernard and Ravenshill, 1995) which are asserted more often than they are demonstrated. There are indications that such arguments are open to question.

First, it is indeed true that Japanese affiliates draw also on local partners, equity and loans and thus the value of their invested capital may underestimate the scale of operations in which they have a leading role. One might counter that unless local participants are seen as mere pawns, such a wider spread of Japanese influence is at the same time a dilution of it. There are also, however, other measures, relating for example to the number of workers employed by Japanese affiliates, or to their output and exports, which are not susceptible to the same criticism.

In 1992, Japanese figures show there were around 1 million

workers employed in manufacturing in all Japanese affiliates in Asia, about 450,000 each in the NIEs and in ASEAN and some 70,000 in China (Abo, 1996, p. 138). At that time there were some 15.7 million workers employed in manufacturing in Japan, 9 million in the NIEs and 14 million in the ASEAN countries (JSY, 1995, pp. 84, 798, 813). Thus those employed in all Japanese affiliates in the NIEs constituted about 5% of the manufacturing workforce and little more than 3% in ASEAN. The total employed, outside Japan, in Asia was equivalent to less than 7% of the manufacturing workforce within Japan. The production anywhere outside Japan, by Japanese affiliated firms, as a share in total Japanese production, was under 5% in 1986 and under 10% in 1995, very much lower than the 20-25% for United States firms or the 15-20% for German firms (*WIR*, 1996, p. 48).

Some have argued that the crucial multiplier of influence and dependence is in the local subcontracting carried out by such affiliates (Ozawa, 1993, Arrighi et al, 1993). Descriptions are given of subcontracting in the automobile industry in Japan, in which over 80% of the value is contributed before the final assembly (Arrighi et al, 1993, p. 51) but significant evidence is not presented about the scale of subcontracting in other industries or in overseas affiliates. In fact many writers complain about the limited extent of local procurement by Japanese invested companies abroad and their heavy reliance on parts from Japan (for example Borrus, 1994) while Japanese investors have shown themselves reluctant to depend on the reliability and quality of local subcontractors (Woronoff, 1984, 268; Kumar and Yuen, 1992, 18; Chiasakul and Silipipat, 1992, 233; Ernst, 1994). Aoki's study of Japanese affiliates in Malaysia indicates that in 1989 less than a quarter of the value of production had been added from local procurement (of materials as well as parts, 1992, p. 81).

Hatch and Yamamura (1996) claim that the growth in exports and production in Asia and the flood of cheap imports into the US have derived largely from Japanese owned subsidiaries or their dependent subcontractors. A simple calculation (not performed by the authors), from figures in the book itself, indicates the fallacy of this. Total exports of manufactures from non-Japan Asia in 1993 were valued at US$459 billion (p. 181). Exports from Japanese affiliates in Asia that year were valued at US$21.6 billion (calculated from pp. 256-57). Exports from Japanese affiliates (which would include the incorporated value of any subcontracted work) were thus worth under 5% of all Asia's manufactured exports. Since 37% of these went to Japan, the use of Asia as a Japanese production platform aimed at international markets seems to have accounted for under 3% of Asia's manufactured exports.

In the second argument, Japanese regional economic power is often ascribed to their technological lead and a claimed dominance over its supply. At the start of 1990s the Japanese were certainly important suppliers of productive machinery to the rapidly industrialising countries of East Asia, both to companies where they had provided investment or loans and, more often, through the open market to others under non Japanese ownership. Yet in no country did they operate without rivals from other sources or account for a majority of machinery imports. In addition, in the NIEs in particular, domestic production was increasing in sophistication and importance. At the start of the 1990s Japan supplied between a quarter and a half of the machinery imports of each of the Asian NIEs and the ASEAN countries, with the highest proportions in 1990 to Thailand, 44%, and to Taiwan, 48% (EIU). There is disagreement as to whether American, European or Japanese companies were relatively more or less willing to transfer the latest or most appropriate technology[8] but there can be little doubt that competition between them, as well as the emergence of new sources within the region, strengthened the hand of the buyers[9].

Through the first half of the 1990s all the East Asian NIEs and the countries of ASEAN experienced continuing industrial growth and most increased their imports of machinery substantially. Japan's role as a supplier did not experience the same precipitate decline as its investment in the region, but neither did it compensate for it. By the mid 1990s Japan's contribution to total machinery imports had fallen somewhat in almost all the countries, only rising in South Korea and remaining constant in Malaysia (EIU).

The belief that Japan holds a stranglehold on high technology in Asia is by now clearly exploded. The way companies in Asia have been successfully climbing the technology ladder is increasingly well documented (Hobday, 1995). Recently, for example, Bennett, the CEO of Intel was quoted as saying that 'Taiwan is fast becoming the world's major design centre for the PC industry'. On the other hand none of Japan's major companies had undergone the structural changes to keep up with the US and he did not rate them as 'world leaders in computing' (*FEER*, 13 May 1999, p. 46).

The third argument is the most far reaching. The increasing role of Japan as overseer and coordinator of a regional division of labour and borderless production, through its network of affiliates in East Asian countries, is much heralded (Bernard and Ravenshill 1995; Tokunaga,

1992a; Morris-Suzuki, 1992). While grandiose plans are easy to cite, the obstacles to operating Just-in-Time production across distances and boundaries are rarely considered and quantitative data is rarely presented.

Dobson, in her analysis of MITI data on overseas affiliates in Asia in 1990, presents little evidence of realisation, as yet. While 55% of all inputs to these firms came from within the country where they were operating and 35% came from Japan, only 11% came from any third country. Of this 11% only a little over a quarter (around 3% in all) came from firms in the same group (1993, p. 48). Dobson herself makes no comment on these figures, but they do present a sobering contrast to both the enthusiastic plans and the doomsday predictions of advocates and Cassandras of Japan's imminent role as brain and nerve centre of the region.

Japan as Saviour in the Crisis?

The Asian recovery got underway in the first half of 1999 (see Table 10.5). There was no evidence, however, that Japanese imports, investment or loans were playing any major role. Total Japanese imports, which had fallen during the crisis, barely changed between the end of 1998 and mid 1999 (although they have been rising since), nor were they refocussing on Asian suppliers. At the onset of the crisis, in the last quarter of 1997, Japan was importing only 11.5% of the exports of the region (not including Taiwan for which quarterly figures are not available). A year later, in the last quarter of 1998, the absolute value of such imports had fallen by over 10% and now represented an even smaller proportion of the region's exports. (*DOTS Quarterly,* September 1999).

In mid 1999, *Far Eastern Economic Review* reported that exports from crisis hit Asian countries 'have started roaring back in recent months' with much of the growth coming from reviving intra regional trade. But as the region began, first slowly and then with accelerating momentum, to pull out of its crisis, Japanese markets did not appear to be playing any leading role. In May 1999, for example, while Singapore's exports to Taiwan grew by 50% over the previous May, those to Japan increased by only 1.5% (1 July 1999, p. 51).

Nor were Japanese loans and investments contributing to the recovery. Although Japanese FDI in the region had become of decreasing importance over the 1990s, Japanese banks had been responsible for much short term lending to the region, directly or as channels for Western funds, and   were also important in triggering the subsequent hot money flight. Because they had so many bad debts at home as well, however, they have been particularly unwilling to write off bad debts (*FEER*, 27 May, 1999, p. 63). In mid 1999 Sender described as a 'flood' the way Japanese companies were 'closing their overseas operations [in Southeast Asia] and fleeing back home, swamped by recession and the refusal of chastened Japanese banks to lend money' (*FEER*, 29 July 1999, pp. 52-54).

Table 10.5 Economic Growth in East and Southeast Asia During the Crisis

	1997	1998[1]	1999[2]
East Asia			
China	8.8	7.8	7.3
Hong Kong	5.0	-5.1	2.0
Taiwan	6.8	4.7	5.3
Southeast Asia			
Indonesia	4.7	-13.2	-0.4
Malaysia	7.5	-7.5	5.2
Philippines	5.2	-0.5	3.0
Singapore	8.9	0.2	5.0
Thailand	-1.3	-10.4	3.7
Japan	-0.7	-2.1	0.9

[1] Estimate [2] Forecast.

Source: *Far Eastern Economic Review*, 13 January, 2000, pp. 72/73.

Conclusion

During the 1990s Japan has been directing increasing proportions of her foreign investment and of her exports towards the countries of East Asia, particularly China. This, hardly surprising, change of strategic orientation, towards the most economically dynamic region in the world, cannot, however be seen as the cause or leading edge of its continuing dynamism nor has it been able to halt the decline in Japan's relative economic influence in the region.

The dramatic fall in the overall value of Japanese foreign investment since 1991 (in a period when world outflows and inflows to East Asia were rising), cut her share dramatically, despite her renewed focus on the region. Intra regional investment far outstripped that of Japan. In trade, the region has been of greater importance to Japan than is Japan to the region. In the 1990s, it is true, Japan's importance as a trading partner for East Asia as a whole has increased (from a low level). This, however, was pretty well all accounted for by the growth of commerce between Japan and China, a growth that nonetheless leaves Japanese influence in China itself very limited compared with the predominance there of the Chinese diaspora (Lever-Tracy et al, 1996).

This chapter has been unable to answer more qualitative questions about indirect dominance, through minority shareholdings and loans, subcontracting and control over the supply of technology and of the coordination of regional production networks. The claims in this respect have, however, been shown to be often undemonstrated and open to question. The numbers employed by Japanese affiliates do not indicate a major presence in the manufacturing sector of the region. Subcontracting may extend this, but there are doubts about how far Japan's elaborate domestic multi layered subcontracting system has actually been transplanted. In a period of rapid industrialisation Japan has been providing less rather than more of the region's machinery imports. Finally the data provided by Dobson indicate that at least as of 1990, schemes for elaborate Japanese intra company coordination of regional production systems were still largely on the drawing board. If such qualitative arguments for Japanese power are to carry a burden of proof counter to the broad evidence of investment and trade figures, they will require much more substantiation.

If Japan had indeed triggered the initial take off of the region and masterminded the 'Asian Miracle', then it might perhaps seem reasonable to presume that any recovery from the crisis would depend on a revival of Japanese initiative, markets and investment. So long as Japan continued to stagnate, there was little to be expected from the capitalism of other parts of Asia which had supposedly now been unmasked as 'ersatz' 'crony capitalism'. Yet again assumptions about who are the only effective historical actors has blocked the questions and analysis that are needed. Thus just as the 'miracle' and the crisis had both taken so many observers by surprise, so did the rapid recovery of the region in 1999. A questioning of the central relevance of Japan to whatever happens in the region seems a necessary precondition to understanding.

Notes

[1]   A survey of the readers of the *Far Eastern Economic Review* found this view widely repeated, with 82% claiming Japan was the most important influence on Asian economies (23 July 1998, p. 23).

[2]   Personal communication by Man-hung Liu, Institute of Modern History, Academia Sinica, Taipei.

[3]   Hong Kong is a financial centre for much of diaspora Chinese capitalism in the region and many Southeast Asian conglomerates have major registered companies and second headquarters here (Lever-Tracy et al, 1996, Ch. 6).

[4]   Redding (1995, 62) calculates that from 1986 to 1991 Japan contributed 29.3% of all FDI to the countries of ASEAN (excluding Singapore) while the Overseas Chinese (here meaning Hong Kong, Taiwan and Singapore) contributed 29.6%.

[5]   Apart from South Korea all these countries have in common the predominance of the Chinese diaspora in their private business and investment activities. The inclusion of South Korea does not affect the picture to any great extent. Korea accounted for under 15% of East Asia's trade and under 9% of its intra regional sources of FDI in 1995 (*WIR*, 1996; DOTS Yearbook, 1995). For discussion and evidence about the Chinese diaspora as an increasingly integrated current within world capitalism see Lever-Tracy et al, 1996 (especially Chapters 5 and 9).

[6]   The proportion to Asia was low throughout the second half of the 1980s. The rise in the 1990s, however, did no more than restore a prior balance. From 1950-1983 the cumulative proportion of notifications directed to Asia had been 26% (Abbeglen and Stalk, 1985, p. 246).

[7]   There was a debate in the mid 1990s on whether the growing proportion of trade flows within East and Northeast Asia was indicative of the emergence of a trade bloc under Japanese dominance. The more convincing analyses indicate that this growth was what would be expected from the growing production and exports of the region, that trade within the supposed bloc was still at a much lower level than that within the EEC or NAFTA and that it was not dominated by Japan (Bobrow et al 1996; Frankel, 1994; Petri, 1994; Saxonhouse, 1994).

[8]   For arguments that contrast favourably the impact of Japan, on the technological capacity of the region, with that of Western countries see Abbeglen and Stalk (1985, pp. 260-262); Aoki (1992, pp. 77/78 and Passim); Arrighi et al (1993); Fujita and Hill (1995, p. 16); Dobson (1993, pp. 45-49); Ozawa (1993). For a negative view of the impact of Japan see for example Bernard and Ravenshill (1995); Morris-Suzuki (1992); Ernst (1994, 190-191); Steven (1990). Borrus (1994) contrasts the recent Western impact favourably with that of Japan.

[9]   Abbeglen and Stalk (1985, p. 242); Borrus (1994, p. 144); Hobday (1995). Fong (1992), using case studies of foreign projects in Malaysia as illustrations, argues that NIE investors in Malaysia introduced more advanced technology than did the Japanese and that they were able to obtain it through their access to German, British and American sources as well as from Japan (pp. 203, 206-216).

References

Abegglen, J. and Stalk, G.Jr. (1985), *Kaisha, The Japanese Corporation*, Basic Books, New York.

Abo, T. (1996), 'The Japanese Production System: the Process of Adaptation to National Settings', in Boyer, R. and Drache, D., *States Against Markets*, Routledge, London.

Ahmad, A., Rao, S.and Barnes, C. (1996), *Foreign Direct Investment and APEC Economic Integration,* Industry Canada, Working Paper 8, February, Toronto.

Aoki, T. (1992), 'Japanese FDI and the Forming of Networks in the Asia-Pacific Region: Experience in Malaysia and its Implications' in Tokunaga, S., *Japan's Foreign Investment and Asian Economic Interdependence:Production, Trade and Financial Systems*, University of Tokyo Press, Tokyo.

Arrighi, G., Ikeda, S. and Irwan, A. (1993), 'The Rise of East Asia: One Miracle or Many?', in Palat, R.A. (ed.), *Pacific-Asia and the Future of the World-System*, Greenwood Press, Westport CT.

Bobrow, D. B., Chan, S. and Reich, S. (1996), 'Southeast Asian Prospects and Realities: American Hopes and Fears', *The Pacific Review*, Vol 9/1.

Borrus, M. (1994), 'Left for Dead: Asian Production Networks and the Revival of US Electronics', in Doherty, E.M., *Japanese Investment in Asia: International Production Strategies in a Rapidly Changing World*, The Asia Foundation, BRIE, Berkeley.

Chiasakul, S. and Silipitat, P. (1992), 'The Role of Japanese Direct Investment in Developing Countries: The Case of Thailand', in *The Role of Japanese Direct Investment in Developing Countries*, Institute of Developing Economies, Tokyo.

Chiu, S.W.K. and Lui, T.L. (1995), 'Hong Kong: Unorganised Industrialism' in Clark, G.L. and Kim, W.B., *Asian NIEs and the Global Economy*, John Hopkins University Press, Baltimore.

Cronin, R. (1991), 'Changing Dynamics of Japan's Interaction with Southeast Asia', *Southeast Asian Affairs*, Institute of Southeast Asian Studies, Singapore.

Dobson, W. (1993), *Japan in East Asia: Trading and Investment Strategies*, Institute of Southeast Asian Studies, Singapore.

*DOTS (Direction of Trade Statistics), Quarterly.*IMF (International Monetary Fund), various dates.

*DOTS (Direction of Trade Statistics), Yearbook.*IMF (International Monetary Fund), various dates.

EIU (Economist Intelligence Unit), *Country Reports*, various dates.

Ernst, D. (1994), *Carriers of Regionalisation: The East Asia Production Networks of Japanese Firms, Berkeley Roundtable on the International Economy*, University of California, Working paper 73, November, Berkeley.

FEER (Far Eastern Economic Review), various dates

Fong, C.O. (1992), 'Foreign Direct Investment in Malaysia: Technology Transfer and Linkages by Japan and Asian NIEs' in Tokunaga, S., *Japan's Foreign Investment and Asian Economic Interdependence: Production, Trade and Financial Systems*, University of Tokyo Press, Tokyo.

Frankel, J. A. (1994), 'Is Japan Creating a Yen Bloc in East Asia and the Pacific', in

Garnaut, R. and Drysdale, P., *Asia Pacific Regionalism*, Harper Educational.

Fujita, K. and Hill, R.C. (1995), 'Global Toyotaism and Local Development', *International Journal of Urban and Regional Research*, 19/1, March.

Gereffi, G. and Hamilton, G. (1996), 'Commodity Chains and Embedded Networks: The Economic Organisation of Global Capitalism', Paper presented at the *Annual Meeting of the American Sociological Association*, New York.

Greenhalgh, S. (1988), 'Families and Networks in Taiwan's Economic Development', in Winckler, E. and Greenhalgh, S. (eds.), *Contending Approaches to the Political Economy of Taiwan*, East Gate, Armonk.

Hamilton, G. (ed.) (1991), *Business Networks and Economic Development in East and Southeast Asia*, Centre for Asian Studies, University of Hong Kong, Hong Kong.

Hatch, W. and Yamamura, K. (1996), *Asia in Japan's Embrace: Building a Regional Production Alliance*, Cambridge University Press, Cambridge.

Hobday, M. (1995), *Innovation in East Asia: The Challenge to Japan*, Edward Elgar, Cheltenham.

Hone, A. (1974), 'Multinational Corporations and Multinational Buying Groups: Their Impact on the Growth of Asia's Exports of Manufactures - Myths and Realities', *World Development*, 2/2, February 1974.

IFS (International Financial Statistics) Yearbook, IMF (International Monetary Fund), various dates

JSY (Japan Statistical Yearbook) Statistics Bureau, Tokyo, various dates.

Kagotani, N. (1998), 'The Overseas Chinese Networks and Wartime Japan: A Japanese Perspective from the 1930s to the Beginning of 1940', Paper presented at the *Ninth Pacific Inter-Congress*, Academia Sinica, Taipei, November.

Kojima, K. and Ozawa, T. (1984), *Japan's General Trading Companies: Merchants of Economic Development*, Development Centre Studies, OECD, Paris.

Kumar, S. and Ng, C. Y. (1992), 'Japanese Manufacturing Investments in Singapore. Linkages with Small and Medium Enterprises' in Institute of Developing Economies, *The Role of Japanese Direct Investment in Developing Countries*, Tokyo.

Lever-Tracy, C., Ip, D. and Tracy, N. (1996), *The Chinese Diaspora and Mainland China, an Emerging Economic Synergy*, Macmillan, Houndsmills and St. Stephens Press, New York.

Lever-Tracy, C. and Tracy, N. (1993), 'The Dragon and the Rising Sun: Market Integration and Economic Rivalry in East and Southeast Asia', *Policy, Organisation and Society*, 6, Summer.

Mitchell, B. and Ravenshill, J. (1995), 'Beyond Product Cycles and Flying Geese: Regionalisation, Hierarchy and the Industrialisation of East Asia', *World Politics*, 47, January.

Morris-Suzuki, T. (1992), 'Japanese Technology and the New International Division of Knowledge in Asia' in Tokunaga, S. (ed.), *Japan's Foreign Investment and Asian Economic Interdependence: Production, Trade and Financial Systems*, University of Tokyo Press, Tokyo.

OECD (1995), *Foreign Direct Investment: OECD Countries and Dynamic Economies of Asia and Latin America*.

Orru, M. (1991), 'The Institutional Logic of Small Firm Economies in Italy and Taiwan', *Studies in Comparative International Development*, Spring, Vol 26/1.

Ozawa, T. (1993) 'Foreign Direct Investment and Structural Transformation: Japan as a

Recycler of Market and Industry', *Business and the Contemporary World*, Spring.

Petri, P. (1994), 'The East Asian Trading Bloc: an Analytical History'. In Garnaut, R. and Drysdale, P., *Asia Pacific Regionalism*, Harper Educational, Pymble.

Redding, G. (1995), 'Overseas Chinese Networks: Understanding the Enigma', *Long Range Planning*, 28/1.

Rob, S. (1990), *Japan's New Imperialism*, Macmillan, Houndmills.

Saxonhouse, G. (1994), 'Trading Blocs and East Asia', in Garnaut R. and Drysdale, P. (eds.) *Asia Pacific Regionalism*, Harper Educational, Pymble.

Tokunaga, S. (1992a), 'Japan's FDI-Promoting Systems and Intra-Asia Networks: New Investment and Trade Systems Created by the Borderless Economy', in Tokunaga, S. (ed.), *Japan's Foreign Investment and Asian Economic Interdependence: Production, Trade and Financial Systems,* University of Tokyo Press, Tokyo.

Tokunaga, S. (1992b), 'Moneyless Direct Investment and Development of Asian Financial Markets: Financial Linkages Between Local Markets and Offshore Centers'. in Tokunaga, S. (ed.), *Japan's Foreign Investment and Asian Economic Interdependence: Production, Trade and Financial Systems*, University of Tokyo Press, Tokyo.

WIR, (World Investment Report, annual), United Nations, various dates.

Woronoff, J. (1984), *Japan's Commercial Empire*, M.E. Sharpe, Armonk.